Technological Change and Organizational Action

Edited by David Preece and Juha Laurila

LONDON AND NEW YORK

This book is dedicated to

Maureen, Jamie and Laura,
and
Eija and Jarkko

First published 2003 by Routledge
11 New Fetter Lane, London EC4P 4EE

Simultaneously published in the USA and Canada
by Routledge
29 West 35th Street, New York, NY 10001

Routledge is an imprint of the Taylor & Francis Group

Typeset in Garamond by Wearset Ltd, Boldon, Tyne and Wear
Printed and bound in Great Britain by Antony Rowe Ltd, Chippenham, Wiltshire

British Library Cataloguing in Publication Data
A catalogue record for this book is available from the British Library

Library of Congress Cataloging in Publication Data
Technological change and organizational action / edited by David Preece and Juha Laurila.
p. cm.
Includes bibliographical references and index.
1. Technology–Management. 2. Technological innovations–Management. I. Preece, David, 1948– II. Laurila, Juha.
T49.5 .T3815 2003
658.4′062–dc21

2002012825

ISBN 0–415–26591–6

Technological Change and Organizational Action

In recent decades an extensive array of changes and innovations have taken place in and across work organizations and networks of organizations and workers, facilitated by new technologies and technological forms. This has initiated an interest in technological change as one of the conditions for organizational action and researchers have begun to draw inspiration from a wider spectrum of conceptual issues, perspectives and theoretical traditions.

This book is interested in both what is occurring at the level of praxis and how this might be understood and theorized. It brings together a comprehensive collection of empirically grounded and theoretically informed research projects which are based upon and deploy new material gathered from studies of organizational practice and which explore a number of technological changes in a variety of organizational, work, sectoral and country contexts. These chapters are informed by contemporary debates within and across theoretical approaches including the sociology of technology, work and organizations, actor network theory, technology as text and metaphor, processual and political perspectives, social and business network-based approaches to the analysis of technology and innovation, and the social construction and shaping of technology.

This book will be essential reading for researchers and advanced students within the fields of technology, work and organizations and also organization studies and management studies.

David Preece is Professor of Technology Management and Organization Studies in the Business School, the University of Teesside. His current research interests include organizational change in the UK public house retailing sector and constructing, using and talking about company intranets.

Juha Laurila is a Senior Research Fellow (The Academy of Finland) in Management and Organization at the Helsinki School of Economics, Finland. His current research interests include institutional and social movement theory and technological change in business firms.

Routledge Studies in Technology, Work and Organizations
Series edited by David Preece

1. Information Society and the Workplace: Spaces, Boundaries and Agency
Edited by Tuula Heiskanen and Jeff Hearne

2. Technological Change and Organizational Action
Edited by David Preece and Juha Laurila

Contents

Illustrations

Figures

Tables

Contributors

Professor Richard Badham, Department of Management, University of Wollongong, Australia.

Dr James Brown, Sir John Cass Business School, London, UK.

Ken Clarke, Gloucestershire Business School, University of Gloucestershire, Cheltenham, UK.

Professor Patrick Dawson, Department of Management Studies, University of Aberdeen, UK.

Professor Hans-Dieter Ganter, Fachhochschule Heilbronn, Germany.

Dr Karin Garrety, Centre for Change Management, University of Wollongong, Australia.

Hannu Hänninen, Helsinki School of Economics and Business Administration, Finland.

Professor Chris Hendry, Sir John Cass Business School, London, UK.

Susanne Hilland, Fachhochschule Heilbronn, Germany.

Professor Oswald Jones, The Business School, Manchester Metropolitan University, UK.

Professor David Knights, Department of Management, Keele University, UK.

Christian Koch, Associate Professor, Institute for Planning, Technical University of Denmark, Lyngby, Denmark.

Dr Juha Laurila, Helsinki School of Economics and Business Administration, Finland.

Dr Darren McCabe, Department of Management, Keele University, UK.

Professor Ian McLoughlin, Newcastle School of Management, University of Newcastle, UK.

Professor David Preece, The Business School, University of Teesside, Middlesbrough, UK.

Sari Yli-Kauhaluoma, Helsinki School of Economics and Business Administration, Finland.

Preface

These are fascinating times for researchers, students and practitioners interested in organizations and technology. There was an emerging awareness during the late 1980s and 1990s that the new forms of information technology, communication media and electronic data interchange, which had been created and developed by hardware and software companies (and, often, subsequently further developed and configured within adopting organizations), offered some challenging and potentially significant implications for the ways in which work might be conducted and organized. In consequence, new insight was called for and provided on the design of jobs, work organization, networking, the experience and meaning of work, and the (changing) nature of organizations, organizing and technology itself.

As the dissemination and awareness of these developments and possibilities gathered momentum during the 1990s, organization studies scholars began to theorize and analyse what was unfolding and what was implied for organizations, the conduct and experience of work, and the actors involved. In particular, researchers in this domain became interested in technological change as one of the conditions for organizational action and began to draw inspiration from a wider spectrum of conceptual perspectives. These include the social construction and social shaping of technology, actor network theory and technology as text and metaphor. By the early 1990s, in addition, more and more attention was being paid to the analysis of technological change as a political process. Such an approach also became the predominant axis around which analysis and discussion was formulated. This situation implied an explicit or implicit opposition to previously dominant technological determinism. Thus, by the mid- to late 1990s, researching and writing about technological change in organization studies was enriched increasingly by a wider variety of perspectives, concepts and theories. This was not unconnected, we suspect, to the new – or even radically new – forms of technology and telecommunications which were simultaneously being introduced. The implications and meanings of the new forms could, perhaps, not be understood without the aid of imported theories or the derivation or development of new means of analysing what was happening at the level of practice. Of course, it took some time for publications to emerge which were based upon a solid grounding in studies and analyses of work and organizational praxis. Moreover, these studies have typically been informed by one of the available or emerging theoretical perspectives, and only in some instances have used an inductive approach to help generate or develop novel theory.

This, it seems to us, is where this area of the organization-oriented field of technology studies is today. The state of affairs reflects the extensive array and variety of changes and innovations, facilitated by new technologies and technological ensembles, which are taking place in and across work organizations and networks of organizations/workers. Only by drawing upon a variety of perspectives and theoretical traditions is there a possibility to gain some purchase upon what is happening.

Our interest in this book is both in what is occurring at the level of praxis, and how this might be understood and theorized. In our view the latter must be grounded in high-quality contemporary empirical research which has, in some way or other, engaged with organizational action and the relevant actors themselves. In particular, we have gathered together research papers which are based upon and deploy new material gathered from studies of organizational practice. These papers also draw upon, and sometimes develop, one or more of the emergent theoretical perspectives relating to technological change and organizational action, while at the same time retaining, where appropriate, the advancements made previously within mainstream management and organization studies. It is important to add the proviso here that we are not implying that we believe recent and current organizational practice is all about change, for we make no such assumption. Indeed, one of the particularly interesting empirical issues relates to continuity and the historical legacy, and the degree to which this influences the present and the emergent.

The chapters in this book reflect an extensive array of empirically grounded and theoretically informed research projects, which explore a number of technological changes in a variety of organizational, work, sectoral and country contexts. Thus the papers we have selected are both empirically rich and theoretically informed by contemporary debates within and, in some cases, across the sociology of technology, work and organizations, actor network theory, technology as text and metaphor, processual/political perspectives, social and business network-based approaches to the analysis of technology and innovation, and the social construction/shaping of technology. In their various ways they are all concerned with understanding and reporting upon what has been happening in organizations at the interface between contemporary technological change, work and the actors involved and affected.

We are confident that this book will be of benefit to researchers and advanced students within the technology, work and organizations field, but will also serve as a source book for more general and other more specific learning and research projects (e.g. knowledge management, modes of control, change agency, regional studies) in organization and management studies more generally. Apart from Chapter 1, the chapters represent a selection of the papers which were originally presented at the Technological Change and Organizational Action subtheme (convened by Juha Laurila and David Preece) of the European Group for Organization Studies Annual Colloquium, held in Helsinki in July 2000.

Acknowledgements

David Preece would like to thank the Business School, University of Portsmouth, where he worked until October 2001, for funding his travel to and attendance at the 2000 EGOS Colloquium in Helsinki; also, along with the Helsinki School of Economics and Business Administration, for contributing towards the costs of a pre-Colloquium visit to Helsinki in 1999, when he and Juha Laurila were in the early stages of planning the subtheme. Particular thanks are due to Kari Lilja for his support and hospitality while David was in Helsinki. Juha would like to thank the Academy of Finland and the Helsinki School of Economics and its foundations for financial support. The EGOS 2000 Colloquium organizers should also be acknowledged for putting together and administering a most enjoyable and academically rewarding conference.

We would also like to thank Drexal Parker of the University of Teesside for his IT expertise in helping to put the final manuscript together, Joe Whiting, the Commissioning Editor for Routledge Research Monographs, and the anonymous reviewers of the original proposal for their constructive observations during the gestation period of this book.

Finally, our appreciation goes to our families – Maureen, Jamie and Laura; Eija and Jarkko – for their understanding and support, and especially for diverting us on occasions from the lonely experience of the word processor.

1 Looking backwards and sideways to see forward

Some notes on technological change and organizational action

Juha Laurila and David Preece

Introduction

Technology in its various forms is an extensively studied phenomenon within the field of management and organization studies. In this chapter we examine actor-focused approaches to technological change. Our main concern is to identify and highlight some of the most prominent work in both this and related streams of literature in order to provide some fruitful starting points for this volume. We recognize that there are numerous ways in which technology has been conceptualized in management and organization studies (see e.g. Roberts and Grabowski 1996; Scott 1998; Preece *et al.* 2000 for reviews). Consequently, it is impossible to do justice to the diversity of this research within this short chapter. We, therefore, aim merely to highlight some of those features which appear to us to be the most significant for understanding why the research field in focus has transpired in its current form.

At the outset, it is necessary to define the focus and background assumptions of this chapter. First, in our theme of technological change and organizational action we remain mainly within the boundaries of contemporary management and organization studies. We do not, therefore, for example, refer to economics-based research traditions which have examined the role of technology in economic growth, its relations to structures of industrial organization and path-dependencies in their development (see e.g. Dosi *et al.* 1992). We also do not draw upon innovation studies, which forms a considerable body of literature within the contemporary social sciences (see e.g. Slappendel 1996). Second, in our conceptualization of technology we accept that it includes both hardware, such as machines or technical devices, and software, such as knowledge and work practices. Such a definition recognizes the open-ended nature of technology and related work practices. We also consider technology to be both socially constructed and as forming part of our lived experience. This means that we do not enter the post-structuralist realm, in the sense that we would then abandon the distinction between action and structure (cf. Reed 1997). Moreover, we consider hardware to be a more elementary form of technology than software, especially in the sense that technology always involves hardware but not necessarily software (cf. Orlikowski 2000). Moreover, while acknowledging that technology is often an outcome product (cf. e.g. Rosenkopf and Tushman 1994), here it is more typically seen as a tool to accomplish organizational tasks. As far as organizational action is concerned, we acknowledge that it takes place simultaneously at various hierarchical levels and functions. This does not mean, though, that we view organizations as isolated entities. Rather, we see them as being in constant interaction with their technical, political, economic and cultural environments, within which technologies and technological change form but one distinct embodiment.

This chapter consists of four sections. Following this brief introduction we next review some of the classical conceptualizations of the nature of the relationship between technological change and organizational action in management and organization studies. We then examine some more recent work on similar issues. We also refer to some recent attempts to develop actor-focused approaches, subsequently extending these frameworks via recent research on organizational networks. Finally, we offer some concluding observations on the current state of actor-focused research on technological change and organizational action.

Looking backwards: the incremental incorporation of organizational actors into the study of technology

Organization-oriented research on technology and technological change has, over the past few decades, only incrementally paid direct attention to the contribution of organizational actors in the construction and shaping of socio-technical configurations or ensembles (see e.g. Bijker *et al.* 1987; Bijker 1995; MacKenzie and Wajcman 1999). Initially, it ignored such matters more or less totally. The classical works here include Thompson (1967) who made a distinction between the so-called technological core of organizations and its peripheral elements. The latter were, in his view, arranged in relation to the degree of complexity, uncertainty and interdependence of the technologies employed in a given organization; in particular he distinguished between pooled, sequential and reciprocal technical processes. The strength of this perspective, and its main legacy for contemporary research in this domain, is that organizational technologies are seen to differ significantly from each other, and, consequently, organizational practices (including, for example, demands for information production and dissemination (Galbraith 1974; Huber 1990)) and the programmability of the technological processes (Sproull and Goodman 1990) also differ.

This structurally determinist stream of research, based on the work of Thompson (1967) and others (e.g. Blauner 1964; Woodward 1965; Perrow 1967), has not produced coherent results on the relationship between technologies and organizations. It has not proved possible to establish clearly stated universal relationships between the characteristics of technologies and organizations, as claims for unilateral relationships between technology and organizations have not held. Common reasons for this include differing operationalizations of the 'technology' concept, and variations in the levels of analysis (Burkhardt and Brass 1990; Roberts and Grabowski 1996). Instead, research findings have been interpreted to show that technologies and organizations are interrelated in a much more complex way, with a number of intermediary dimensions (Meyer *et al.* 1993), including industry (e.g. Barley 1986) and country-specific ones (e.g. Lincoln *et al.* 1986; Sorge 1991).

Early actor-focused studies and their recent extensions

Until the early 1980s, the mainstream of research concentrating on the relationship between technology and organizations thus suffered from conceptual weaknesses and paid too little attention to the contexts in which technologies were being applied. It also paid insufficient attention to the role of organizational actors in technological change. The emergence of actor-focused approaches on technologies and technological change (Pettigrew 1973; Buchanan and Boddy 1983; Wilkinson 1983; Weiss and Birnbaum 1989) was related to the emergence of a strategic choice perspective on organizations (Child 1972). Also influential were social action approaches within the sociology of organizations (e.g. Goldthorpe *et al.* 1968), and labour process approaches (Braverman 1974; Noble 1979, 1984) which looked at how technologies were constructed and used for certain political purposes. From these positions, it was quite elementary to incorporate actors into the concepts and models which were developed to analyse technological change and organizations.

The above-mentioned studies drew attention to such matters and issues as socio-economic contexts, legitimacy and the interests of the organizational actors shaping and guiding technological change. The cognitive orientations, skills and professional background of managerial actors were shown to influence the content of technological change (Daft 1978). Technological change came to be seen as an outcome of political negotiation and practices (e.g. Kanter 1988; Barley 1990; see also Clausen *et al.* 2000). Others (e.g. Burgelman and Sayles 1986; Day 1994; Laurila 1998) argued that the mobilization of collective co-operation is necessary, especially in producing radical change and innovation in the core technologies of organizations. Thus, in addition to political manoeuvring, emotional involvement and voluntary co-operation are implicated in (especially) major technological change. Further, managerial succession can be an important source of actor-driven technological change (Child and Smith 1987; Langley and Truax 1994; Boeker 1997; Zucker and Darby 1997).

Actor-focused studies have thus shown that people are central to the adoption (Preece 1995) and implementation of technological change. This is notwithstanding the fact that managerial actors do not always learn from their experiences, but tend to repeat previously used recipes for managing technology (Pennings *et al.* 1994; Tyre and Orlikowski 1994). What is more, managers have been found to oppose new technologies which destroy their existing competencies (Tushman and Anderson 1986) or to make obsolete previous technologies to which they are emotionally committed (Burgelman 1994; Stuart and Podolny 1996). None the less, organizational actors make a difference, as they decide on which forms of new technology to introduce (Laurila 1997) or the specific ways to implement externally initiated technological change (Boisot 1995; Argyres 1996). But how do actors actually influence technological change? Help in moving forward in this direction is at hand from those co-evolutionary approaches on technological change and organizational action which were first developed at the ecological and organizational levels of analysis.

Recent co-evolutionary and structurational approaches

The above-mentioned approaches are similar in that they consider either technology or organizational actors to be the main source of change. In contrast, co-evolutionary approaches have emphasized the ever-enduring interplay of technologies and organizations over time. At the ecological level, the work by Michael Tushman and his collaborators on technological discontinuities has been influential (Tushman and Anderson 1986; Anderson and Tushman 1990; Rosenkopf and Tushman 1994). This research has shown how technological change is produced by and influences organizational communities formed around technological innovations. Specific technologies and the relevant organizational communities have their own developmental trajectories which are, at specific phases, connected to each other. A distinctive characteristic of these approaches is the emphasis placed upon the cyclicality of change (cf. Van de Ven and Poole 1995). Technological and organizational change, it is argued, occurs in the form of successive cycles including continuous and discontinuous periods. Technological change is an outcome both of intentional action and serendipity. The origins of technological change reside both inside and outside individual organizations. Most importantly for research on technological and organizational change in general, however, these studies problematize not only the relationship between technology and organizations, but also its nature as a continuous and mutually entwined process.

A number of other researchers (e.g. Child 1997; Zucker and Darby 1997) have also had an interest in the cyclical change of technologies at the inter- and intra-organizational levels. These approaches assume that not only do people make choices between alternative technologies, but also that their interests and the competencies involved in making these choices alter during the course of the change process. Organizational actors learn while gathering and applying experiences from a specific technology. Conceptualizations of the evolutionary change of technologies at the intra-organizational level (e.g. Burgelman 1996; Lovas and Ghoshal 2000) have shown that actors at different levels of the organizational hierarchy continuously initiate change. The mechanisms through which these initiatives are processed, however, differ between individual organizations and over time.

A number of recent studies take the position that technological change and organizational action should be examined as an inherently inseparable and incrementally unfolding phenomenon (e.g. Barley 1986; DeSanctis and Poole 1994; McLoughlin and Dawson, Chapter 2, this volume). Orlikowski (1992, 2000) distinguishes between what she labels the 'core' and 'tangential' characteristics of technology and, on the other hand, technology as an artefact and technology in use. Here, technology is seen as a behavioural product which is constructed in use, influenced largely by the characteristics and interests of the user. Thus, in contrast to technology as text or metaphor perspectives (see below), for example (Grint and Woolgar 1997), the core of technology as a materially factual phenomenon is preserved.

Co-evolutionary approaches demonstrate that technologies and organizations evolve in continuously changing and unanticipated ways (cf. e.g. Hänninen, Chapter 6, this volume). Technological discontinuities are typically considered to be an outcome of intentional collective action, while periods of incremental technological change encourage organizational actors to adapt themselves to current technologies. None the less, organizational action always includes unanticipated consequences and opportunities for innovation that cannot be attributed to technologies per se. There is also, of course, an implication that technological change is only partly manageable by organizational actors.

Looking sideways: recent actor and network-focused approaches to technological change

On the basis of the above review and discussion it is clear that, despite substantive findings and conceptual development, the previous work in the technology and organizations domain is problematic. In this present section, we attempt to build upon this literature by elaborating and extending it with respect to actor conceptualizations. We begin by examining some of the more recent developments in actor-focused approaches within the field of management and organization studies. In the subsequent section we extend this by drawing upon some of the developments in network-based approaches to technological change.

Recent developments in actor-focused approaches to technological and organizational change

Why is it important to take an actor-focused approach to the analysis of technological change? According to Floyd and Lane (2000) this is mainly because previously developed technologies or technological competencies cannot be relied upon. Hence there is an impetus for continuous development and improvement, which makes relevant actors' change capabilities critical. What is more, technological change rarely takes place in specified locations, such as product development departments. Rather, changes occur at various locations within and outside the organization as a result of the actions (spoken and unspoken) of a variety of people. The actors involved in the adoption and introduction of technologies thus change throughout the life-cycle of these technologies.

A number of models have been developed to help us understand the social and political processes through which organizational actors produce technological change (see e.g. Burgelman and Sayles 1986; Clark *et al.* 1988; Day 1994; Preece 1995; Laurila 1998; see also Dutton *et al.* (1997) and Floyd and Wooldridge (1997) for examples of more general models of actor-based change). Perhaps the key message of these studies is that organizational action oriented towards change is highly susceptible to contextual and situational circumstances (Ashford *et al.* 1998). At the same time, it cannot be assumed that organizational actors are eager to initiate change, because of potential threats to their reputation and credibility. This is especially the case for middle managers, who must 'read the wind' (Dutton *et al.* 1997) in order to assess the prospects of their change initiatives succeeding in specific organizational settings. Of course, change is not necessarily objected to by other actors, who, among other considerations, may feel that a change initiative provides relief from existing responsibilities (Cox 1997).

Thus the shaping of technology by organizational actors cannot be understood irrespective of the contexts in which they operate. The nature and extent of trust and other socio-economic relations between actors is especially important here. This is especially so when we acknowledge that actors' social identities are created in part by the specificities of the technologies they use and promote (Kilduff *et al.* 1997). Technologies thus help make organizational actors who they are (Covaleski *et al.* 1998) while, at the same time, people construct their identities through technology (Preece and Clarke, Chapter 3, this volume). A key message of studies of and writings on organizational identity for research on technological change is that organizational action emerges not only from within the pre-structured world within which people are located, but also from the socially constructed identities which give meaning to social action.

Without contending that social identities are constant or internally coherent (cf. e.g. Golden-Biddle and Rao 1997), it may be argued that changes in the social identities of actors are a prerequisite for them being active initiators of change (Gioia and Thomas 1996). This is especially the case if one takes the view that there exists a close relationship between action and the actors' interpretations of that action (e.g. Daft and Weick 1984). It follows that organizational actors' interpretations of particular technologies change during the course of the events which unfold when that technology is implemented (Barr 1998). Moreover, social interaction which triggers or accelerates the emergence of new interpretations may also facilitate technological change. However, it has been argued that the social identities of actors change incrementally at most, with only a loose coupling between identities and presentation of self. That is, various rhetorical practices are employed to create representations of what is actually being done (Fine 1996); for example, different types of linguistic expression may be used at different phases of technological change (cf. Sillince 1999).

There is a certain resonance between the above mentioned work and that of sociologists who have taken up a 'Technology as Text and Metaphor' (TTM) position with respect to the conceptualization and understanding of technology and technological change. TTM takes up an anti-essentialist position, often influenced by Actor-Network Theory (ANT) (Callon 1986; Latour 1987; Law 1991) but wishing to move beyond it (see Grint and Woolgar 1997). With TTM, 'the boundary between the social and the technical is part of the phenomenon to be investigated' rather than being taken as given, as with essentialist perspectives, and 'The nature and characteristics of technical capacity, what the machine can and cannot do, what it is for, how it can be enrolled and controlled and so on are crucial matters for the sociologist' (Grint and Woolgar 1997: 37). A 'given' technology can be understood, or 'read', only in the particular social and organizational contexts in which it is found, and different representations of the technology and organization can be created and found via different metaphors or texts. This is, of course, to take up a strong relativist position, perhaps nowhere better illustrated than in the 'Woolgar and Grint vs. Kling' journal exchange about 'Guns and Roses' and the relative chances of being shot by one or the other (see Kling 1991a, 1991b, 1992a, 1992b; Woolgar 1991; Woolgar and Grint 1991, and, for a summary of and commentary upon the debate, McLoughlin 1997). We concur with McLoughlin (1999) and McLoughlin and Dawson (Chapter 2, this volume) on this debate and TTM more generally: although the debate is ultimately inconclusive it shows that TTM fails to capture the materiality of technology as a 'hard place'. Actor-Network Theory does attempt to preserve this materiality, but does so in a way which neglects to adequately recognize and theorize the influence of internal and external (to the organization) socio-economic and political contexts upon the emerging and evolving nature, form and representations of technology.

Despite the problematics of TTM and ANT approaches, we none the less believe that they have helped to invigorate the field of technological change and organizational action in recent years, not least because of the ways in which they have challenged 'traditional wisdom', as outlined in particular in the first main section. We did state earlier that we would restrict ourselves to a critical overview of research on technology in management and organization studies domains. Clearly, in referring to TTM and ANT theorizing, we have moved away from this by 'looking sideways'. Further, it is necessary to continue to do this in order to indicate how the actor in technology creation and shaping and technological change can be foregrounded while at the same time preserving the facticity of technology. For us, the most thoroughgoing way in which this has been achieved to date is through social construction (SCT) and socio-economic shaping (SST) of technology perspectives, with the important caveat that there is built in a stronger dose of political behaviour in the 'social'. McLoughlin and Dawson (this volume) have expressed the challenge graphically as being 'to provide analytical means which enable us to make sense of the materiality of technology as a "hard place" and the interpretative flexibility of technology as a social construct'.

The SCT perspective draws upon the sociology of scientific knowledge, preserving the materiality of technology while at the same time analysing the ways in which and the conditions under which actors create socio-technical configurations which come to be 'stabilized' (cf. to TTM, where it is argued there is to be no stabilization) as consensus emerges over a design option (see e.g. Pinch and Bijker 1987; Bijker 1995). More recently, it has been argued that, insofar as stabilization can be said to have occurred, it could equally likely be through the exercise of power in the form of the imposition of a particular configuration or 'ensemble' by a dominant group of actors (Blosch and Preece 2000; Koch 2000). Stabilization should always be seen as pro tem, i.e. not implying that this configuration is fixed for ever (see also Orlikowski 2000). Or as McLoughlin and Dawson (this volume) express the matter, 'we do not have to reject the concept of stabilization . . . but need to render it as a more contingent, malleable and iterative conceptualization of the obduracy of technology'. Design objectives are by no means purely 'technically rational', and can also or instead be economic, financial, social and/or moral.

SST perspectives focus, in the main, upon what becomes of a particular socio-technical ensemble once it has passed the original conception stage. In essence, the argument is that technology is shaped by the economic, technical, political, gender and social circumstances in which it is designed, developed and utilized (MacKenzie and Wajcman 1999). Winner (1977, 1980) argues that the perspective has been informed by a concern to democratize technological decision-making, or, at least, subject it to social accountability and control. The work by McLaughlin *et al.* (1999) has much in common with the social shaping perspective, focusing as it does upon the ways in which 'local', in-use configurations of the socio-technical ensemble is the way in which it comes, if at all, to be valuable to the organization through processes of de- and re-stabilization. Local work and organizational knowledge is seen as being just as important as knowledge about the technology per se, and the ensemble becomes highly specific to the particular organization or work locale into which it is introduced and deployed.

Drawing together some of the key messages for research on technological change which have emerged from actor-focused perspectives, one can say that intensive interaction among organizational actors is a key precursor and facilitator of technological change. This is in particular because of the social identities which await to be transformed in the process of such change. There are a number of matters, however, which have not been adequately addressed to date. For example, although some attention has been given to political conflicts between the actors involved in technological change (e.g. Knights and Murray 1994), internal contradictions in the identities of actors have been largely ignored (but see Hayes and Walsham 2000). Moreover, whereas the intra-organizational ecology of technological change (e.g. Burgelman and Sayles 1986; Laurila 1998) has been conceptualized in a preliminary manner, the conceptualization of organizational actors in these models lacks sophistication.

The further development of our understanding of issues and phenomena such as the above presents a significant challenge because it requires an at least partial abandonment of the previously dominant organization-level unit of analysis. A more sophisticated understanding of the relationship between organizational action and technological change demands that more attention be paid to the connections between actors across formal organizational boundaries (cf. e.g. Yli-Kauhaluoma, Chapter 8, this volume), and, indeed, between formal organizations and individualized 'home'/'tele'-workers (see Jackson and Van der Wielen 1998; Jackson 1999; Felstead and Jewson 2000). We know little, for example, about the nature of the relationship between managers (and others) and competitor organizations, customers and (especially technology) suppliers, and about how specific change proposals are created in this interaction (but see Koch, Chapter 4, this volume). It is thus appropriate to move to a consideration of network-based approaches to the analysis of technological change.

Social network approaches to technological change

Network perspectives on technological change within management and organization studies have become more common in recent years. We are able here to refer to only a small part of this large and increasingly diverse literature (see Oliver and Ebers 1998).

Perhaps the most important message for technological change research emerging out of network-based studies is that being part of a social network is likely to increase the innovativeness of an individual organization (see e.g. Powell *et al.* 1996; Kraatz 1998). Although there are some opposing views (see e.g. Robertson *et al.* 1996), the argument seems plausible as long as the innovativeness concept refers to susceptibility for change, without contending that these 'innovative organizations' are necessarily more 'up-to-date' than others. If this is the case, then it would appear that multiple connections between an organization and its environment yield more change initiatives than do few connections. Matters become even more complex, however, when we problematize the nature of the change which has emerged out of organizational networks. For example, technological change may be original or imitative, focused or dispersed, and so forth. Studies have examined whether weak ties to a large number of actors (Granovetter 1973) or strong ties to a limited number of actors (Uzzi 1996) within a network are more likely to produce change. While this debate continues, there is some evidence to show that technological change may be promoted most efficiently by weak network ties when information in itself (Nagarajan and Mitchell 1998) and the search for suitable technological partners (Gulati 1999) is the most critical issue for the change in question. On the other hand, when the originality of the technological/organizational innovations involved is the key issue, intensive co-operation among a small number of actors seems to be more influential than non-intensive co-operation among a large number of actors (Uzzi 1996; Kraatz 1998).

It is worth noting that network studies such as those referred to above have not fully problematized the issue of organizational knowledge. Researchers who have focused on the different forms of knowledge and obstacles to the transfer of knowledge between organizational actors (e.g. Szulanski 1996; Lam 1997) have shown how difficult such knowledge transfer may be or become. Other researchers (e.g. Hargadon and Sutton 1997) have problematized the distinctive competencies of actors who specialize in transferring knowledge across organizational boundaries. Moreover, some work (e.g. Browning *et al.* 1995; Newell *et al.* 2001) has shown how knowledge flows between organizational actors may be disrupted significantly by intra- and extra-organizational contingencies. When, on the other hand, organizational networks as sources of information are stressed, it is necessary to be aware of the possibility of 'bandwagon effects' (Abrahamson 1996) in the emergence of a specific dominant design of technology within an organizational domain (Tegarden *et al.* 1999). Studies such as these have shown that adaptability rather than anticipation is crucial in adjusting individual organizations to technological change.

The more sophisticated conceptualizations of organizational networks have not only examined firms connected to each other through supplier and customer relations, but also via different institutional layers, such as professional associations (Swan *et al.* 1999a, 1999b). These studies have shown that the forms taken by technological change are in part dependent upon the societal and institutional structures in which organizational networks are embedded. What is more, organizational networks weaken and strengthen at different stages of technological change (Madhavan *et al.* 1998), and individual members of a network may modify the structure of the network (for example, early adopters of a new technology may improve their power position at the expense of others (Burkhardt and Brass 1990)).

To summarize, network approaches extend the more traditional organization-level approaches to technological change by paying attention to the diversity of institutional and social layers within which an organization is embedded. The incorporation of such approaches into the wider body of research on technological and organizational change, however, is at an early stage. The problematics of network studies need to be addressed as this work develops. For example, although numerous complexities of and dimensions in social networks are recognized, typically studies draw on only a few of them (cf. Osborn and Hagedoorn 1997; Hardy and Phillips 1998).

Network studies have usually concentrated on technical material or information flows as a basis for identifying a given network. In contrast, more symbolic or personal relations-based networks (Geletkanycz and Hambrick 1997) have been largely omitted. It may be anticipated that different network forms will follow different paths in their development, and thus the conceptualization of networks should take these differences into account. Moreover, network studies suffer from a bias towards recognizing only the positive effects of social networks while ignoring the negative ones. Hardy and Phillips (1998) found that being part of a network can be counterproductive for an individual organization. Finally, it is important to note that a key challenge for technological change research adopting a network perspective is that to date networks have rarely been examined in the context of a specific change initiative, such as, indeed, technological change (Kraatz 1998; Oliver and Ebers 1998).

Trying to see forward: some conclusions

The main concern of this chapter has been to critically review actor-focused research on technology in the management and organization studies field, and hence to position this volume in relation to that diverse stream of literature. Not surprisingly, our examination shows that technology in this domain has been studied from a variety of perspectives and for a variety of purposes. Only occasionally does one find an attempt to elaborate on the relationship between technology and organizations in any comprehensive way (e.g. Tushman and Rosenkopf 1992; McLoughlin 1999; Orlikowski 2000; McLoughlin and Dawson, Chapter 2, this volume). More typically, studies concentrate only on certain specific dimensions of the technology–organizations relationship, and vice versa (cf. Griffith 1999). In part, such approaches reflect an awareness of the contextual and situation-specific nature of technology and organizations. We continue to need these focused studies, as the concepts and perspectives through which technology and organizations are examined have their own life-cycle; in part because of continuing innovations and developments at the level of practice in the technology/actor/organization/network interface, new or amended concepts and perspectives which give purchase upon these new forms are always required. As an example, a recent study by Argyres (1999) found that new information technologies may enable co-ordination between firms in previously unspecified ways.

The derivation of new concepts and perspectives, however, would appear to present some significant challenges for a number of writers, for, as Barley (1996) has observed, much contemporary research in the domain uses concepts developed through empirical study several decades ago. For example, although co-evolutionary approaches on technological change may seem dominant, it is not difficult to find studies which approach technology and organizations from a strong social determinist perspective (e.g. Grint and Woolgar 1997; Zucker and Darby 1997; Dyerson and Mueller 1999). Zucker and Darby (1997), for example, show how technological championing, recruitment and transformative leadership may produce an 'ethos of change' in an established firm. On the other hand, and even more often, contemporary studies (e.g. Heijltjes 2000; Spell 2001) adopt a broadly technological determinist perspective. In a similar vein, a deterministic position is taken on the relationship between technology and organizational performance by Klassen and Whybark (1999), and on the flexibility demands of new technology by Lei *et al.* (1996).

Much of the recent writing on the so-called 'virtual organization' (e.g. Davidow and Malone 1992; Barnatt 1995; Grenier and Mates 1995) has strong technologically determinist undertones, where new ICT, satellite and telecommunications technology (especially where allied to networked forms of working and collaboration) are theorized in such a way that it becomes difficult if not impossible to differentiate between 'the people' and 'the technology' (Jackson 1996); in other words, our gaze is directed to information capture, processing, storage, dissemination and, in some writing (e.g. Barnatt 1995) to the immersion of the subject (people) into this 'virtual world' of cyberspace. Quite apart from the major problems which arise here of conceptualizing and understanding the nature and lived experience of this 'virtual reality' (see e.g. Castells 1996, and his contrasting concept of 'real virtuality'; also Castells 2000), along with all the well-known deficiencies of the technological determinist position, there is a real danger that the 'technical' becomes fused with the 'social' in such a way that the role, influence and reflective experience of human actors is lost or concealed.

The diversity of approaches within the technology and organizations literature is due not only to the different epistemological and methodological positions adopted by researchers, but also to their different objectives and concerns. On the one hand, for example, we have researchers who aim to present 'practitioner-friendly' findings on connections between technology and organizations. MacGrath *et al.* (1992) and Tegarden *et al.* (1999), for example, examined how firms may anticipate technological change and increase their adaptability for such change in competitive environments. On the other hand, we find studies which are concerned to understand why and how specific forms of technology are adopted in specific institutional and historical contexts (e.g. Sorge 1991), or how specific ways of using technology become institutionalized (e.g. Karnoe and Garud 1997; Zyglidopoulos 1999). One reason why the latter studies are particularly valuable is because they encourage us to recognize the 'taken-for-grantedness' of technology; that is, technology is examined as both a technical and a socio-cultural phenomenon.

We thus anticipate that technology will remain a key focus of interest for future research and writing within the field of management and organization studies. Why do we have this confidence? One reason is that this research is vital not only for those who recognize the centrality of technology to contemporary and emerging work and organizations, but also for those who do not have technology and organizations as a key point of interest. This does not come as a surprise where technology is defined very broadly to incorporate (for example) incentive and surveillance systems (e.g. Kilduff *et al.* 1997). In addition, technologically sophisticated firms or industries provide interesting settings for studies using a variety of theoretical perspectives (cf. Sewell and Wilkinson 1992; Pliskin *et al.* 1997; Taylor and Bain 1999). While there are other recent examples of how theorization in this domain may be deployed as a means of enlightening other fields of research (see e.g. Caspar 2000; Brocklehurst 2001), it may be anticipated that this form of 'inter-disciplinary research' or 'borrowing' will continue into the future. This is not least, again, because of the continuing technological and organizational innovation and change which has been initiated, developed and disposed of by people in their various communities of practice, and societal, institutional and organizational settings and contexts.

References

Abrahamson, E. (1996) 'Management fashion', *Academy of Management Review* 21: 254–285.

Anderson, P. and Tushman, M. (1990) 'Technological discontinuities and dominant designs: a cyclical model of technological change', *Administrative Science Quarterly* 35: 224–241.

Argyres, N. (1996) 'Capabilities, technological diversification and divisionalization', *Strategic Management Journal* 17: 395–410.

—— (1999) 'The impact of information technology on coordination: evidence from the B-2 Stealth bomber', *Organization Science* 10, 2: 162–180.

Ashford, S.J., Rothbard, N.P., Piderit, S.K. and Dutton, J.E. (1998) 'Out on a limb: the role of context and impression management in selling gender-equity issues', *Administrative Science Quarterly* 43: 23–57.

Barley, S.R. (1986) 'Technology as an occasion for structuring: evidence from observation of CT scanners and the social order of radiology departments', *Administrative Science Quarterly* 31, 1: 78–108.

—— (1990) 'The alignment of technology and structure through roles and networks', *Administrative Science Quarterly* 35, 1: 61–103.

—— (1996) 'Technicians in the workplace: ethnographic evidence for bringing work into organization studies', *Administrative Science Quarterly* 41: 404–441.

Barnatt, C. (1995) *Cyberbusiness: Mindsets for a Wired Age*, Chichester: Wiley.

Barr, P.S. (1998) 'Adapting to unfamiliar environmental events: a look at the evolution of interpretation and its role in strategic change', *Organization Science* 9, 6: 644–669.

Bijker, W. (1995) *Of Bicycles, Bakelites and Bulbs: Toward a Theory of Socio-Technical Change*, Cambridge, MA: MIT Press.

Bijker, W., Hughes, T. and Pinch, T. (eds) (1987) *The Social Construction of Technological*

Systems: New Directions in the Sociology of History and Technology, Cambridge, MA: MIT Press.

Blauner, R. (1964) *Alienation and Freedom*, Chicago: University of Chicago Press.

Blosch, M. and Preece, D. (2000) 'Framing work through a socio-technical ensemble: the case of Butler Co.', *Technology Analysis and Strategic Management* 12, 1: 91–102.

Boeker, W. (1997) 'Executive migration and strategic change: the effect of top manager movement on product-market entry', *Administrative Science Quarterly* 42: 213–236.

Boisot, M. (1995) 'Is your firm a creative destroyer? Competitive learning and knowledge flows in the technological strategies of firms', *Research Policy* 24: 489–506.

Braverman, H. (1974) *Labour and Monopoly Capital*, New York: Monthly Review Press.

Brocklehurst, M. (2001) 'Power, identity and new technology homework: implications for "new forms" of organizing', *Organization Studies* 22, 3: 445–466.

Browning, L.D., Beyer, J.M. and Shetler, J.C. (1995) 'Building cooperation in a competitive industry: sematech and the semiconductor industry', *Academy of Management Journal* 38, 1: 113–151.

Buchanan, D. and Boddy, D. (1983) *Organizations in the Computer Age*, Aldershot: Gower.

Burgelman, R.A. (1994) 'Fading memories: a process theory of strategic business exit in dynamic environments', *Administrative Science Quarterly* 39, 1: 24–56.

—— (1996) 'A process model of strategic business exit: implications for an evolutionary perspective on strategy', *Strategic Management Journal* 17, Summer Special Issue: 193–214.

Burgelman, R.A. and Sayles, L.R. (1986) *Inside Corporate Innovation: Strategy, Structure and Managerial Skills*, New York: Free Press.

Burkhardt, M.E. and Brass, D.J. (1990) 'Changing patterns or patterns of change: the effects of a change in technology on social network structure and power', *Administrative Science Quarterly* 35, 1: 104–127.

Callon, M. (1986) 'The sociology of an actor-network: the case of the electric vehicle', in M. Callon, J. Law and A. Rip (eds) *Mapping the Dynamics of Science and Technology: Sociology of Science in the Real World*, Basingstoke: Macmillan.

Caspar, S. (2000) 'Institutional adaptiveness, technology policy, and the diffusion of new business models: the case of German biotechnology', *Organization Studies* 21, 5: 887–914.

Castells, M. (1996) *The Information Age: Economy, Society and Culture, Volume 1: The Rise of the Network Society*, Oxford: Blackwell.

—— (2000) *The Information Age: Economy, Society and Culture, Volume 1: The Rise of the Network Society* (2nd edn), Oxford: Blackwell.

Child, J. (1972) 'Organization structure, environment and performance: the role of strategic choice', *Sociology* 6, 1: 2–22.

—— (1997) 'Strategic choice in the analysis of action, structure, organizations and environment: retrospect and prospect', *Organization Studies* 18: 43–76.

Child, J. and Smith, C. (1987) 'The context and process of or-ganiza-tional transformation: Cadbury Limited in its sector', *Journal of Management Studies* 24, 6: 565–593.

Clark, J., McLoughlin, I., Rose, H. and King, R. (1988) *The Process of Technological Change*, Cambridge: Cambridge University Press.

Clausen, C., Dawson, P. and Nielsen, K. (2000) 'Political processes in management, organization and the social shaping of technology', *Technology Analysis and Strategic Management* 12: 5–15.

Covaleski, M.A., Dirsmith, M.W., Heian, J.B. and Samuel, S. (1998) 'The calculated and the avowed: techniques of discipline and struggles over identity in Big Six public accounting firms', *Administrative Science Quarterly* 43: 293–327.

Cox, J.R.W. (1997) 'Manufacturing the past: loss and absence in organizational change', *Organization Studies* 18, 4: 623–654.

Daft, R.L. (1978) 'A dual core model of organizational innovation', *Academy of Management Journal* 21, 2: 193–210.

Daft, R.L. and Weick, K.E. (1984) 'Toward a model of organizations as interpretation systems', *Academy of Management Review* 9, 2: 284–295.

Davidow, W. and Malone, M. (1992) *The Virtual Corporation*, New York: Harper Business.

Day, D.L. (1994) 'Raising radicals: different processes for championing innovative corporate ventures', *Organization Science* 5, 2: 148–172.

DeSanctis, G. and Poole, M.S. (1994) 'Capturing the complexity in advanced technology use: adaptive structuration theory', *Organization Science* 5, 2: 121–147.

Dosi, G., Giannetti, R. and Toninelli, P.A. (eds) (1992) *Technology and Enterprise in a Historical Perspective*, Oxford: Clarendon Press.

Dutton, J.E., Ashford, S.J., O'Neill, R.M., Hayes, E. and Wierba, E.E. (1997) 'Reading the wind: how middle managers assess the context for selling issues to top managers', *Strategic Management Journal* 18, 5: 407–425.

Dyerson, R. and Mueller, F. (1999) 'Learning, teamwork and appropriability: managing technological change in the department of social security', *Journal of Management Studies* 36, 5: 629–652.

Felstead, A. and Jewson, N. (2000) *In Work, at Home: Towards an Understanding of Homeworking,* London: Routledge.

Fine, G.A. (1996) 'Justifying work: occupational rhetorics as resources in restaurant kitchens', *Administrative Science Quarterly* 41: 90–115.

Floyd, S. and Wooldridge, B. (1997) 'Middle management's strategic influence and organizational performance', *Journal of Management Studies* 34, 3: 465–485.

Floyd, S.W. and Lane, P.J. (2000) 'Strategizing throughout the organization: managing role conflict in strategic renewal', *Academy of Management Review* 25, 1: 154–177.

Galbraith, J. (1974) 'Organization design: an information processing perspective', *Interfaces* 4: 28–36.

Geletkanycz, M.A. and Hambrick, D.C. (1997) 'The external ties of top executives: implications for strategic choice and performance', *Administrative Science Quarterly* 42: 654–681.

Gioia, D.A. and Thomas, J.B. (1996) 'Identity, image, and issue interpretation: sensemaking during strategic change in academia', *Administrative Science Quarterly* 41: 370–403.

Golden-Biddle, K. and Rao, H. (1997) 'Breaches in the boardroom: organizational identity and conflicts of commitment in a nonprofit organization', *Organization Science* 8, 6: 593–611.

Goldthorpe, J., Lockwood, D., Bechhofer, F. and Platt, J. (1968) *The Affluent Worker*, Cambridge: Cambridge University Press.

Granovetter, M.S. (1973) 'The strength of weak ties', *American Journal of Sociology* 78: 1360–1380.

Grenier, R. and Mates, G. (1995) *Going Virtual: Moving Your Organisation into the 21st Century*, New York: Prentice Hall.

Griffith, T.L. (1999) 'Technology features as triggers for sensemaking', *Academy of Management Review* 24, 3: 472–488.

Grint, K. and Woolgar, S. (1997) *The Machine at Work: Technology, Work and Organization*, Cambridge: Polity Press.

Gulati, R. (1999) 'Network location and learning: the influence of network resources and firm capabilities on alliance formation', *Strategic Management Journal* 20: 397–420.

Hardy, C. and Phillips, N. (1998) 'Strategies of engagement: lessons from the critical examination of collaboration and conflict in an interorganizational domain', *Organization Science* 9, 2: 217–230.

Hargadon, A. and Sutton, R.I. (1997) 'Technology brokering and innovation in a product development firm', *Administrative Science Quarterly* 42: 716–749.

Hayes, N. and Walsham, G. (2000) 'Competing interpretations of computer-supported cooperative work in organisation contexts', *Organization* 7: 49–67.

Heijltjes, M.G. (2000) 'Advanced manufacturing technologies and HRM policies: findings from chemical and food and drink companies in the Netherlands and Great Britain', *Organization Studies* 21, 4: 775–805.

Huber, G. (1990) 'A theory of the effects of advanced information technologies on organizational design, intelligence and decision making', *Academy of Management Review* 15: 47–71.

Jackson, P. (1996) 'The virtual society and the end of organisation', Working Paper, Uxbridge: Department of Management Studies, Brunel University.

Jackson, P. (ed.) (1999) *Virtual Working: Social and Organisational Dynamics*, London: Routledge.

Jackson, P. and Van der Wielen, J. (eds) (1998) *Teleworking: International Perspectives*, London: Routledge.

Kanter, R.M. (1988) 'When a thousand flowers bloom: structural, collective and social conditions for innovation in organization', *Research in Organizational Behavior* 10: 169–211.

Karnoe, P. and Garud, R. (1997) 'Path creation and dependence in the Danish wind turbine field', in J. Porac and M. Ventresca (eds) *Social Construction of Industries and Markets*, Oxford: Pergamon.

Kilduff, M., Funk, J.L. and Mehra, A. (1997) 'Engineering identity in a Japanese factory', *Organization Science* 8, 6: 579–592.

Klassen, R.D. and Whybark, D.C. (1999) 'The impact of environmental technologies on manufacturing performance', *Academy of Management Journal* 42, 6: 599–615.

Kling, R. (1991a) 'Computerisation and social transformations', *Science, Technology and Human Values* 16: 342–367.

—— (1991b) 'Reply to Woolgar and Grint: a preview', *Science, Technology and Human Values* 16: 379–381.

—— (1992a) 'Audiences, narratives, and human values in social studies of technology', *Science, Technology and Human Values* 17: 349–365.

—— (1992b) 'When gunfire shatters bone: reducing socio-technical systems to social relationships', *Science, Technology and Human Values* 17: 381–385.

Knights, D. and Murray, F. (1994) *Managers Divided: Organisation Politics and Information Technology Management*, Chichester: Wiley.

Koch, C. (2000) 'The ventriloquist's dummy? The role of technology in political processes', *Technology Analysis and Strategic Management* 12: 119–138.

Kraatz, M. (1998) 'Learning by association? Interorganizational networks and adaptation to environmental change', *Academy of Management Journal* 41: 621–643.

Lam, A. (1997) 'Embedded firms, embedded knowledge: problems of collaboration and knowledge transfer in global cooperative ventures', *Organization Studies* 18, 6: 973–996.

Langley, A. and Truax, J. (1994) 'A process study of new technology adoption in smaller manufacturing firms', *Journal of Management Studies* 31, 5: 619–652.

Latour, B. (1987) *Science in Action: How to Follow Scientists and Engineers Through Society*, Milton Keynes: Open University Press.

Laurila, J. (1997) 'The thin line between advanced and conventional new technology: a case study on paper industry management', *Journal of Management Studies* 34, 2: 219–239.

—— (1998) *Managing Technological Discontinuities: The Case of the Finnish Paper Industry*, London: Routledge.

Law, J. (ed.) (1991) *A Sociology of Monsters: Essays on Power, Technology and Domination*, London: Routledge.

Lei, D., Hitt, M.A. and Goldhar, J.D. (1996) 'Advanced manufacturing technology: organizational design and strategic flexibility', *Organization Studies* 17, 3: 501–523.

Lincoln, J., Hanada, M. and McBride, K. (1986) 'Organizational structures in Japanese and American manufacturing', *Administrative Science Quarterly* 31: 33–364.

Lovas, B. and Ghoshal, S. (2000) 'Strategy as guided evolution', *Strategic Management Journal* 21, 8: 875–896.

MacGrath, R.G., MacMillan, I.C. and Tushman, M.L. (1992) 'The role of executive team actions in shaping dominant designs: towards the strategic shaping of technological progress', *Strategic Management Journal* 13, winter: 137–161.

MacKenzie, D. and Wajcman, J. (eds) (1999) *The Social Shaping of Technology* (2nd edn), Buckingham: Open University Press.

McLaughlin, J., Rosen, P., Skinner, D. and Webster, J. (1999) *Valuing Technology: Organisations, Culture and Change*, London: Routledge.

McLoughlin, I. (1999) *Creative Technological Change: The Shaping of Technology and Organisations*, London: Routledge.

McLoughlin, I. (1997) 'Babies, bathwater, guns and roses', in I. McLoughlin and M. Harris (eds) *Innovation, Organizational Change and Technology*, London: International Thomson Press.

Madhavan, R., Koka, B.R. and Prescott, J.E. (1998) 'Networks in transition: how industry events (re)shape interfirm relationships', *Strategic Management Journal* 19: 439–459.

Meyer, J., Tsui, A. and Hinings, C.R. (1993) 'Configurational approaches to organizational analysis', *Academy of Management Journal* 36: 1175–1195.

Nagarajan, A. and Mitchell, W. (1998) 'Evolutionary diffusion: internal and external methods used to acquire encompassing, complementary, and incremental technological changes in the lithotripsy industry', *Strategic Management Journal* 19: 1063–1077.

Newell, S., Scarbrough, H. and Swan, J. (2001) 'From global knowledge management to internal electronic fences: contradictory outcomes of intranet development', *British Journal of Management* 12: 95–111.

Noble, D. (1979) 'Social choice in machine design', in A. Zimbalist (ed.) *Case Studies in the Labour Process*, New York: Monthly Review Press.

—— (1984) *Forces of Production: A Social History of Industrial Automation*, New York: Alfred A. Knopf.

Oliver, A.L. and Ebers, M. (1998) 'Networking network studies: an analysis of conceptual configurations in the study of inter-organizational relationships', *Organization Studies* 19, 4: 549–583.

Orlikowski, W.J. (1992) 'The duality of technology: rethinking the concept of technology in organizations', *Organization Science* 3, 3: 398–427.

—— (2000) 'Using technology and constituting structures: a practice lens for studying technology in organizations', *Organization Science* 11, 4: 404–428.

Osborn, R.N. and Hagedoorn, J. (1997) 'The institutionalization and evolutionary dynamics of interorganizational alliances and networks', *Academy of Management Journal* 40, 2: 261–278.

Pennings, J., Barkema, H. and Douma, S. (1994) 'Organizational learning and diversification', *Academy of Management Journal* 37, 3: 608–640.

Perrow, C. (1967) 'A framework for the comparative analysis of organizations', *American Sociological Review* 32: 194–208.

Pettigrew, A.M. (1973) *The Politics of Organizational Decision Making*, London: Tavistock.

Pinch, T. and Bijker, W. (1987) 'The social construction of facts and artifacts: or how the sociology of science and the sociology of technology might benefit each other', in W. Bijker, T. Hughes and T. Pinch (eds) *The Social Construction of Technological Systems: New Directions in the Sociology of History and Technology*, Cambridge, MA: MIT Press.

Pliskin, N., Romm, C. and Markey, R. (1997) 'E-mail as a weapon in an industrial dispute', *New Technology, Work and Employment* 12: 3–12.

Powell, W.W., Koput, K.W. and Smith-Doerr, L. (1996) 'Interorganizational collaboration and the locus of innovation: networks of learning in biotechnology', *Administrative Science Quarterly* 41: 116–145.

Preece, D.A. (1995) *Organizations and Technical Change: Strategy, Objectives and Involvement*, London: Routledge.

Preece, D.A., McLoughlin, I. and Dawson, P. (eds) (2000) *Technology, Organizations and Innovation: Critical Perspectives on Business and Management*, London: Routledge.

Reed, M.I. (1997) 'In praise of duality and dualism: rethinking agency and structure in organizational analysis', *Organization Studies* 18, 1: 21–42.

Roberts, K. and Grabowski, M. (1996) 'Organizations, technology and structuring', in S. Clegg, C. Hardy and W. Nord (eds) *Handbook of Organization Studies*, Thousand Oaks, CA: Sage.

Robertson, M., Swan, J. and Newell, S. (1996) 'The role of networks in the diffusion of technological innovation', *Journal of Management Studies* 33, 3: 333–359.

Rosenkopf, L. and Tushman, M. (1994) 'The coevolution of technology and organization', in J. Baum, and J. Singh (eds) *Evolutionary Dynamics of Organizations*, Oxford: Oxford University Press.

Scott, W.R. (1998) *Organizations: Rational, Natural and Open Systems* (4th edn), Upper Saddle River, NJ: Prentice Hall.

Sewell, G. and Wilkinson, B. (1992) '"Someone to watch over me": surveillance, discipline and the just-in-time process', *Sociology* 26, 2: 271–289.

Sillince, J.A.A. (1999) 'The role of political language forms and language coherence in the organizational change process', *Organization Studies* 20, 3: 485–518.

Slappendel, C. (1996) 'Perspectives on innovation in organizations', *Organization Studies* 17, 1: 107–129.

Sorge, A. (1991) 'Strategic fit and the societal effect: interpreting cross-national comparisons of technology, organization and human resources', *Organization Studies* 12, 2: 161–190.

Spell, C. (2001) 'Organizational technologies and human resource management', *Human Relations* 54, 2: 193–213.

Sproull, L. and Goodman, P. (1990) 'Technology and organizations: integration and opportunities', in P. Goodman and L. Sproull (eds) *Technology and Organizations*, San Francisco: Jossey-Bass.

Stuart, T. and Podolny, J. (1996) 'Local search and the evolution of technological capabilities', *Strategic Management Journal* 17, 1: 21–38.

Swan, J., Newell, S. and Robertson, M. (1999a) 'National differences in the diffusion and design of technological innovation: the role of inter-organizational networks', *British Journal of Management* 10: 45–59.

—— (1999b) 'Central agencies in the diffusion and design of technology: a comparison of the UK and Sweden', *Organization Studies* 20, 6: 905–931.

Szulanski, G. (1996) 'Exploring internal stickiness: impediments to the transfer of best practice within the firm', *Strategic Management Journal* 17, Winter Special Issue: 27–43.

Taylor, P. and Bain, P. (1999) ' "An assembly line in the head": work and employee relations in the call centre', *Industrial Relations Journal* 30: 101–116.

Tegarden, L.F., Hatfield, D.E. and Echols, A.E. (1999) 'Doomed from the start: what is the value of selecting a future dominant design?', *Strategic Management Journal* 20: 495–518.

Thompson, J.D. (1967) *Organizations in Action*, New York: McGraw-Hill.

Tushman, M.L. and Anderson, P. (1986) 'Technological discontinuities and organizational environments', *Administrative Science Quarterly* 31, 3: 439–465.

Tushman, M.L. and Rosenkopf, L. (1992) 'Organizational determinants of technological change: towards a sociology of technological evolution', *Research in Organizational Behavior* 14: 311–347.

Tyre, M.J. and Orlikowski, W.J. (1994) 'Windows of opportunity: temporal patterns of technological adaptation in organizations', *Organization Science* 5, 1: 98–118.

Uzzi, B.D. (1996) 'The sources and consequences of embeddedness for the economic performance of organizations: the network effect', *American Sociological Review* 61: 674–698.

Van de Ven, A.H. and Poole, M.S. (1995) 'Explaining development and change in organizations', *Academy of Management Review* 20: 510–540.

Weiss, A.R. and Birnbaum, P.H. (1989) 'Technological infrastructure and the implementation of technological strategies', *Management Science* 35, 8: 1014–1026.

Wilkinson, B. (1983) *The Shopfloor Politics of New Technology*, London: Heinemann.

Winner, L. (1977) *Autonomous Technology*, Cambridge, MA: MIT Press.

—— (1980) 'Do artifacts have politics?', *Daedalus* 109: 121–136.

Woodward, J. (1965) *Industrial Organization: Theory and Practice*, New York: Oxford University Press.

Woolgar, S. (1991) 'The turn to technology in social studies of science', *Science, Technology and Human Values* 16: 20–50.

Woolgar, S. and Grint, K. (1991) 'Computers and the transformation of social analysis', *Science, Technology and Human Values* 16: 368–378.

Zucker, L.G. and Darby, M.R. (1997) 'Present at the biotechnological revolution: transformation of technological identity for a large incumbent pharmaceutical firm', *Research Policy* 26: 429–446.

Zyglidopoulos, S. (1999) 'Initial environmental conditions and technological change', *Journal of Management Studies* 36, 2: 241–262.

2 The mutual shaping of technology and organisation

'Between Cinema and a Hard Place'[1]

Ian McLoughlin and Patrick Dawson

'Hard place' = 'the tangible, evocative materials and forms that make up our world'
'Cinema' = 'the way in which the world is mediated through language and imagery'

Introduction

Over the past twenty years, academics have generated a vast body of work exploring the interactions between computer-based technology and organisation (see e.g. Preece *et al.* 2000). However, there is little consensus over the effects of these technologies. As has been observed recently, the organisational consequences of adoption are diverse and contradictory while the content of the technology itself appears to be a 'moving target' (witness the rapid emergence of internet applications and electronic business) rendering attempts to identify its effects even more complex (Robey 2000). At a conceptual level, many recent reviews covering theoretical developments in this area also leave the strong impression of a field which is incomplete, confused and contested (see e.g. Fleck and Howells 1997; McLoughlin 1999; Roberts and Grabowski 1996). As Orlikowski has argued, there is a need for a 'fundamental re-examination' of the core concept of 'technology' if our understanding of its interaction with organisation is to be advanced (1992: 398).

The main objective of this chapter is to contribute to such a re-examination. Our basic argument is that most theoretical models in organisational analysis have largely failed to deal adequately with the conceptualisation of 'technology' (computer or otherwise) and its relationship to organisational 'variables'. A recurrent problem has been the explanatory role of technology as a material artefact/system which has 'effects' on organisation. The early work on the technology–organisation relationship, conducted in the wake of the first large-scale applications of computing technology in the 1950s, identified 'technology' as a significant – even determining – variable in organisational analysis (Dawson 1996: 46–49; Scarbrough and Corbett 1992). This position has of course been subject to substantial revision and critique, in particular, from more critical vantage points where social and political processes within organisations have been emphasised (Knights and Murray 1994; MacKenzie and Wajcman 2000). While endorsing many of these arguments, we contend here that these still do not adequately resolve the 'technology problem'.

In order to provide more purchase on this issue we endorse the view that new theoretical resources from outside of mainstream management and organisation theory need to deployed to 'open up' – in analytical terms – the 'black box' of technology. As we will argue, the notion of technology as a social construct is extremely useful but this leaves open a number of questions concerning the role, if any, of the materiality of technology. At the same time, when 'technology' is re-introduced into any explanatory framework there is a danger that its capabilities and characteristics are seen purely as intrinsic rather than as social constructs (Russell and Williams 2000: 44). We seek to develop, therefore, the notion of the 'mutual' shaping of technology and organisation where both can be understood as an outcome of intertwined social and material processes. This occurs as technologies are configured and re-configured during their production and consumption across shifting organisational contexts of development, supply, adoption, use, redevelopment and so on. Technologies in this way are also represented through a shifting narrative of language and image which give 'meaning' to them in particular interpretative arenas. The challenge in developing our understanding of the organisation–technology relationship is to provide the analytical means which enable us to make sense of the analytical space between the materiality of technology as 'hard place' and the interpretative flexibility of technology as a social construct – 'cinema'.

Re-examining and redefining technology

If a reconstruction of the concept of technology is needed in order to improve our understanding of the developing technology–organisation relationship, what broad options do we have in the conceptualisation and definition of technology in order to make sense of its effects on organisations? Grint and Woolgar (1997: 10) make a useful distinction between approaches to defining technology in terms that delimit between its 'non-human' and 'human' elements and those which seek to recast the relationship between technology and humans 'as a network rather than as parallel but separate systems'. The first approach, we would suggest, typifies much of the post-Woodward work in organisational analysis that has sought to explore the technology–organisation relationship. The second approach has emerged from work in the sociology of technology. This is having an increasing – although still in our view limited – influence on organisational studies.

Where 'technology' is considered (as in some sense) separate from 'organisation' a key issue is where the boundary between the two should be drawn. We would suggest that the broader or more expansive the definition of 'technology' the more significant an explanatory variable or category it becomes. The narrower or more restrictive the definition the more the independent effects of 'technology' become relatively insignificant, or not significant at all, when compared to the influence of other 'social' factors. The 'expansiveness' or 'restrictiveness' of the definition of technology may be given in two ways (Orlikowski 1992: 398–399). First, in terms of the 'scope' of the definition of technology – that is, what is defined as technology – and thereby in Grint and Woolgar's terms the boundary between the 'human' and the 'non-human'. Second, in terms of the 'role' ascribed to technology – that is, the nature of the interaction between 'technology' and human and organisational variables – and thereby the attribution of 'independent effects' to technology itself.

The more expansive the definition of technology the more its *scope* is defined in broad terms as embodying not just physical 'hardware' but also 'social technologies' in the form of the tasks, techniques and 'know-how' needed by humans to use 'hardware' and its *role* as having a strong determining influence on organisational structure and behaviour. The more restrictive the definition of technology the more its *scope* is limited to 'hardware' and its *role* represented as a 'softer' determinism or 'reference point' whereby technical effects are more or less strongly 'mediated' during the shaping of organisation by human agents (Orlikowski 1992: 401–402). The position is complicated where restrictive definitions of the scope of technology are combined with expansive definitions of its role; that is, representations of machines as almost causing reflex behaviours on the part of humans in a very crude form of technological determinism. Similarly, it is possible to adopt more expansive definitions of the scope of technology but to associate these with 'soft' determinism.

In the pioneering work of the 1950s and 1960s the scope of technology was, initially at least, defined in restrictive terms as physical artefacts or systems. However, its role was frequently seen to have apparently direct, readily observable and seemingly unmediated effects on organisational behaviour (see e.g. Sayles 1958). As the sophistication of the field advanced, some of these early theorists developed more expansive definitions of the scope of technology by, for instance, incorporating knowledge of how and what to use technology for. Woodward's (1970: 4) definition of 'production systems' as 'the collection of plant, machines, tools and recipes available at a given time for the execution of the production task and the rationale underlying their utilisation' is one of the classic attempts to do this. It is also the case that not all analysts saw the role of technology as unmediated. Woodward (1980: 72) again, although often associated with her apparently deterministic conclusion that, 'there is a particular form of organisation most appropriate to each technical situation', also saw the role of 'technology' as being mediated, to some extent, by the interpretative action of management in realising the need to adapt organisational structures to the requirements of the production system. Indeed even in these early studies, often accused of a crude technological determinism, the notion of 'technology' being associated with variable 'organisational outcomes' was sometimes recognised. For instance, Rosemary Stewart (1971: 228), in her work on the effects of computerisation on management, observed that it was 'misleading to generalise about the impact of the computer on management'. However, the failure to explain such variations in terms other than problems of organisational adaptation to some form of technological requirements has meant that much of this early work has inevitably been labelled as 'technological determinist' and thereby a critical reference point for nearly all subsequent organisational theory and research on the topic.

The critique of technological determinism emerged as a core concern of more critical organisational analysis in the 1970s. One of the strongest original attacks on technological determinism came in the seminal contribution of Braverman (1974) and the 'labour process theory' which it spawned (see e.g. Knights *et al.* 1985; Knights and Wilmott 1988; Thompson 1989, and for recent reviews Parker 1999; Smith and Thompson 1998). However, more mainstream organisational theorists also sought to give primacy to social choices in organisational and work design over and above technological requirements (Child 1972; Davis and Taylor 1973) while other critics highlighted the political nature of decision-making concerning the adoption and deployment of new technology (see e.g. Pettigrew 1973).

These contributions provide the basis for what has been variously termed the organisational politics/process or power-process analysis of technological change (Badham 1993; Thomas 1994). This approach has been developed in a variety of empirical studies of new computer-based technologies spanning the past twenty years (see e.g. Baldry 1988; Buchanan and Boddy 1983; Clark *et al.* 1988; Clausen and Koch 1999; Dawson *et al.* 2000; McLoughlin and Clark 1994; Preece 1995; Salzman and Rosenthal 1994; Thomas 1994; Walsham 1993; Wilkinson 1983; Willcocks *et al.* 1996). Unlike labour process theory, which in its early formulations at least seemed to be replacing one form of determinism (technological) with another (economic), the politics/process perspective requires viewing such contingent factors as no more than contextual *referents* for decision-makers when designing work and organisations around new technology. At the same time, these decisions are the product of, and subject to, the influence of competing interests within organisations as sub-groups and other coalitions seek to defend and advance their positions.

For example, Wilkinson (1983) argued that the introduction of new technology in organisations is best conceived as a *process* with indeterminate outcomes. The organisational outcomes of technological change are therefore the result of more or less unique organisational processes of change, in particular at workplace level. It was here, within the adopting context, that management, unions and workforce – not the external impact of a uniform external force 'technology' – shape the substantive outcomes of change. The resultant form of technology and work organisation should be viewed, not primarily as a reflection of the technical capabilities and characteristics of production technology, but 'as an outcome which has been chosen and negotiated within adopting organisations' (Wilkinson 1983: 20).

One consequence of adopting the politics/process perspective is that the search for an explanation of the effects of technological change moves rapidly from concerns over the scope and role of technology to an attempt to understand the full context, substance and process of change and the manner in which these are shaped by intra-organisational political and cultural processes (Dawson 1994). In a number of cases, it is not what the technology can do which is analytically important for understanding its organisational effects, but rather context-dependent organisational decisions concerning its deployment and use. In order to avoid technological determinism, then, the analytical definition of technology itself disappears as a worthy analytical question. Definitions of the scope of technology in this literature are in the main restrictive in character and thereby limited to references to apparatus, equipment, hardware and software. By the same token any 'effects' that technology has are seen to be highly mediated by organisational processes and to have only very weak mediating influence as an 'intervening variable' on such processes. Indeed, in the light of this, it has been suggested that 'to consider the impact of a particular technology is to consider the wrong question, or at best to consider only part of the issue' (Huczynski and Buchanan 1991: 276). In effect 'technology' is effectively 'squeezed' as a 'variable' out of the analytical frame of organisational research.

Nevertheless, some analysts in this tradition have made a case for its rehabilitation as one factor capable of 'independent influence' of specific aspects of work and its organisation at particular points in the process of change. The point, as Clark *et al.* (1988) sought to argue, was not to 'throw the technology baby out with the determinist bathwater'. Accordingly they suggested that the scope of 'technologies' could be defined 'restrictively' but in more sophisticated analytical terms as 'engineering systems' comprising an 'architecture' or set of general design and engineering principles and an 'implementation' or realisation of this architecture in particular material technologies. In addition, these general characteristics could be further defined in the adopting context in terms of detailed design and realisation in particular material ways such as the layout of equipment, its physical appearance, aesthetics, ergonomics and so on. The role of 'technology' so construed, it was argued, could be seen as a form of 'soft determinism' both enabling and constraining influences on detailed choice and negotiation in relation to organisational issues such as task requirements, job content, work organisation and to some extent the control of work.

In a highly influential contribution to this debate Zuboff (1988) makes a key distinction between the 'automating' and 'informating' characteristics and capabilities of computer-based technologies. The former capability enables the accomplishment of tasks previously undertaken by human intervention and the latter the electronic representation of organisational functioning in a manner which makes human activity far more visible. Her point though, in line with the politics/process approach, is that how these capabilities are exploited is a matter of organisational choice concerning the deployment and use of these systems. The tendency that she observes in the application of these systems to exploit their automating and 'panoptic' potential is, Zuboff contends, not a technical inevitability; rather, this reflects the particular social choices made in specific organisational contexts, the cumulative consequence of which is to act as a constraint on the potential for organisational and social transformation which, she contends, is also embodied in these 'smart' technologies.

But does such an analysis transcend technological determinism? Grint and Woolgar (1997) claim not. They argue that while recognising that such effects are a product of the way technologies are used, these kinds of arguments still tend to view the capacities of information and computing technologies as an 'intrinsic' feature of these systems. Thus Zuboff, for instance, appears to argue that organisations are free to choose to use computing and information technologies in ways which automate rather than informate, but they are unable to challenge the informating capacities of the technology itself. Thus 'the impression given is that there can be no dispute over the potential capacity of the technology, just whether or not this (actual) potential has been realised' (Grint and Woolgar 1997: 134). Furthermore, this gives the strong impression that it is the characteristics and capabilities of advancing 'computer technology' which are transformative, at least where humans are innovative enough to reshape organisations to permit these possibilities to be translated into virtual forms of work and organisational practice. However, this is to treat such technology as though it is a separate and parallel system to the social and organisational.

Beyond technological determinism: technology as a social construct

To move beyond technological determinism in the analysis of the technology–organisation relationship, Grint and Woolgar argue for a consideration of the insights provided by work in the sociology of technology. The starting point here is the social constructivist observation that 'technology does not have any influence which can be gauged independently of human interpretation' (Grint and Woolgar 1997: 10). In this sense what a technology is and is not, and what it can and cannot do, are all socially constructed:

> Technologies, in other words, are not transparent; their character is not given; and they do not contain an essence independent of the nexus of social actions of which they are a part. They do not 'by themselves' tell us what they are or what they are capable of. Instead, capabilities – what, for example, a machine will do – are attributed to the machine by humans.
>
> (Grint and Woolgar 1997: 10)

Thus we are invited to consider that there is no boundary between the technical (non-human) and the social (human) *other than that* which is socially defined. As such, 'our knowledge of technology is in this sense essentially social; it is a construction rather than a reflection of the machine's capabilities' (Grint and Woolgar 1997: 10). Once this is accepted, they argue, 'technologies' should no longer be treated as 'purely technological', but rather must be seen as the outcome of a process of social construction. As Bijker and Law put it, technologies are 'heterogeneous' and as such, 'embody trade-offs and compromises' in the form of 'social, political, psychological, economic, and professional commitments, skills, prejudices, possibilities, and constraints' (1992: 10).

A key concept in the sociology of technology is the notion of 'interpretative flexibility'; that is, what technologies are and what they can do is in principle open to a wide range of possible explanations. As such, the development of technology may be regarded as a multidirectional process where a range of alternative design options exist and are gradually eliminated in the innovation process. 'Technology' takes an obdurate form when a common set of understandings and interpretations is established among relevant social groups in relation to what it can and cannot do, and so forth (Pinch and Bijker 1987: 28). In this sense, 'technology' is socially constructed through a process of 'closure'. This excludes alternative interpretations and understandings being attributed to the machine or system and enables a dominant set of understandings to prevail – at this point the technology is said to have 'stabilised'.

There has been considerable debate within the sociology of technology concerning the value of these notions of 'closure' and 'stabilisation'. Indeed, some critics suggest that this line of argument ultimately runs close to conventional formulations of a technology being capable of having independent 'effects' and 'impacts' once social shaping is dealt with (see Grint and Woolgar 1997). One alternative position that has emerged suggests that what a technology can and cannot do can be more accurately viewed as the product of continual representation and re-representation in the interpretative contexts in which they are being produced and consumed. There is therefore no 'stabilisation' or 'closure' that produces a seemingly 'obdurate' technology. The proposition is that technology is best seen as if it were a 'text' which is being continually 'written' and 'rewritten' as it is produced and consumed (see e.g. Grint and Woolgar 1997). In these terms 'technology' is no more than 'congealed social relations' (Latour 1991) as particular understandings of technology establish some kind of transitory primacy.

Until recently, the 'interpretative flexibility' of technology has been a proposition that has remained largely outside of the analytical gaze of organisational researchers. In part this has reflected a view that, while technologies may be socially shaped in some way, the social processes involved take place largely, if not exclusively, outside of the adopting organisational context (Clark *et al.* 1988; Wilkinson 1983). However, more profoundly it has reflected a theoretical reluctance and conceptual incapacity to 'open the black box' of technology to reveal its own foundations in organisational and broader socio-economic relationships. As we have noted above, the preferred option has been to seek to downplay the analytical significance of technology altogether in an attempt to avoid the determinism of earlier studies.

Having said this, how readily might our understanding of the technology–organisation relationship be advanced by viewing technology as a social construct? The explanatory and practical benefits of seeking to open the black box of technology in this way are by no means always clear-cut. For example, the celebrated debate between Woolgar/Grint and Robert Kling provides an example of the kind of ultimately unproductive polarisation that debate on this issue is prone to produce (see Grint and Woolgar (1992, 1997); Kling (1991a, 1991b, 1992a, 1992b); McLoughlin (1997); and Woolgar and Grint (1991) for a summary and discussion). In a challenge to at least some social constructivist arguments, Kling contended that material technologies – such as computer systems – must be permitted some transformative power within any social analysis of change in contemporary organisations and society. Woolgar and Grint sought to illustrate the latent 'determinism' of such a view by subjecting it to the acid 'test' of illustrating how the apparently 'hard' material fact that guns have the transformative capacity to kill and maim is in the end a social construction. Ultimately, that a gun has the capacity to kill requires the uniting in a particular network of the technical artefact with a set of interpretative frameworks that provide this particular 'reading' of technical capacity.

Such debates highlight the problems that lay in the analytical space between the 'hard place' of technology and the 'interpretative' social worlds in which meanings and understandings of technology are constructed. According to Russell and Williams (2000: 44), for example, we are left with questions such as: how to conceptualise the 'malleability' or 'negotiability' of technology; what role to attribute to the materiality of technology in rendering systems 'durable and reproducible in space'; whether we can attribute to artefacts 'effects' over and above the 'socially constructed properties attributed to them'; and how and in what ways social interactions and communications might be technologically mediated (e.g. as in the increasing electronic mediation of work).

In our view, if the insights of the sociology of technology are to be incorporated into the analysis of the technology–organisation relationship then the issue of the materiality of technology must be confronted. The concepts of 'closure' and 'stabilisation' appear to offer some means of understanding the social construction of technology as resulting in artefacts and systems which have a degree of obduracy or 'epistimological hardness'. However, recent debates in the sociology of technology have tended to emphasise the 'interpretative flexibility' of technology as a continual and continuing process. The problem with this is that we are led into a world where 'technology' appears to reside only in language (McLaughlin *et al.* 1999:

226). There is a danger that in this world analysis becomes a matter of (albeit sometimes entertaining) word games to show the apparently infinitesimal number of ways of constructing the meaning of technology that makes the apparent material certainty that, for example, guns kill, a socially constructed absurdity. Indeed, followed to extremes it seems we are obliged to banish from our conceptual language the notion of technology having 'effects' of any kind on or within organisations, and in some quarters exhorted to find new language which will rid us of this dangerous thought altogether (Grint and Woolgar 1997: 114–115).

So far we have shown that despite many efforts to avoid, counter or embody notions of technological determinism, this determinist concern has been difficult to move beyond in any satisfactory manner. It has remained a recurrent analytical problem, even with the increasing theoretical sophistication of the analytical tools at our disposal to help us avoid such a position. In the next section, we seek to move this debate further by developing a 'mutual shaping' perspective.

Bringing materiality back home: the mutual shaping of technology and organisation

In terms of everyday life within organisations, there remains a strong sense – indeed an increasing one – in which technology as a material force shapes or influences the nature of that experience. As Kumar (1995: 15) observes, it would be 'foolhardy to deny' the kinds of advances in information and computing technology which appear to have transformed the world of work and organisations and much more. By the same token, when theorising these transitions and trends, it would also seem to make little sense to deny that 'technology' has some influence over the process and outcomes of change. One challenge, as we would see it, is to bring the materiality of our everyday experience of technology 'back in' to our analytical frameworks. However, this has to be done in a way that does not represent this 'inescapable part' of everyday organisational 'reality' in a 'partial and one-sided way' (Kumar 1995: 34). We need also to understand and show how what we take to be 'technology' came to be 'technology' in the first place. It is only in the context of this accomplishment that we can legitimately talk of 'technology' – as shorthand for a complex of social and technical relationships – having 'effects', again fully recognising that these 'effects' are contingent, context dependent and not necessarily universal in character.

An important contribution to developing our understanding of these terms is provided by the socio-economic shaping perspective which is concerned to identify the dynamic and interactive form, content and socio-economic context of technological development (see Williams (2000) for a recent review). In this approach technological innovations, initially shaped in one organisational context, are configured into working technologies by local customisation post-adoption (Badham 1995). In turn, if a line of technological development is to endure, the experience and knowledge gained by users in this way needs to be embodied in future innovations by suppliers within their own organisations (Williams 1997). Once again, these consequent innovations are subject to local configuration in adopting contexts and so on. The possibilities of post-adoption innovation by users and its reappropriation as re-innovation by suppliers leads to a view of technology as a highly malleable socio-technical phenomenon which is given meaning in particular and ever-changing contingent circumstances (Fleck 1993).

We would agree that seen in these terms the idea of technologies becoming 'stabilised' in organisations, especially where this is seen as being brought about by some kind of consensus between social groups, does not capture the dynamic and iterative nature of the processes involved both within adopting organisations, and also in terms of relationships with supplier, customer and other relevant organisations. However, this does not mean that we have to reject the concept of 'stabilisation'. This concept (and associated notions of how some degree of shared understanding is reached about a technology) needs to be adapted to suit the circumstances of technology development, deployment and use found in work organisations. In a very helpful contribution to this line of development, McLaughlin *et al.* (1999) have focused on how 'stabilised' technologies brought into an organisation via some external design/development process can be 'destabilised' in adoption and, potentially at least, 'restabilised' in use within organisations (McLaughlin *et al.* 1999). The key here is to focus not only on the way technology is socially shaped in its production but also on how it is socially shaped through its consumption. In this latter respect, McLaughlin *et al.* argue that it is through the manner and the extent to which a technology comes to be valued in use within an organisation that we can see both processes of 'destabilisation' – as a generic system is customised to fit local circumstances – but also 'restabilisation' – where post-adoption innovation by users embeds a technology within a particular organisational context and culture (McLaughlin *et al.* 1999: 227).

Critically, the extent to which the closure and stabilisation occurs in such processes is variable and does not concur with the notion of 'closure through consensus' which marks 'stabilisation' in the sociology of technology. In a similar line of argument McLoughlin *et al.* (2000) suggest that technologies become stabilised through a relative congruence between the belief systems of competing interest groups within organisations – an inherently unstable situation in some circumstances where competing belief systems exist, but a far more obdurate one where power-holding groups are able to 'manage meaning' and provide legitimacy to one particular set of meanings and understandings. In similar vein, Blosch and Preece (2000) suggest that stabilisation through 'consensus' occurs only in settings where power-holding groups within organisations are able to manage meaning in such ways as to 'manufacture such consent', not least by restricting access to forums where competing definitions of technology are resolved. In a variation on this theme, other writers have sought to show how the processes of stabilisation/destabilisation/restabilisation of technology can act to provide a focus for the building of coalitions of interest in organisations – for example, Koch (2000 and Chapter 4, this volume) shows how Enterprise Resource Planning (ERP) systems can enable coalitions between hitherto disparate organisational functions whose interpretative worlds become 'socially glued' together through attempts to develop shared interpretations and understanding of the technologies' capabilities and use.

The concept of technology as a type of 'boundary object' has also been used in several recent studies to weaken even further the notion that closure obtained through consensus is the principal means through which technologies become stabilised with fixed meanings. For example, Garrety and Badham (2000) suggest that technologies become stable where there is sufficient common understanding to enable these groups to combine in various ways to try to design, adopt, implement and/or use them and yet where the technology may mean different things to different social groups. Similarly, McLaughlin *et al.* (1999) suggest that the effective production of technologies (in the sense of their being adopted and implemented in organisations) may well require a process of 'destabilisation' to enable local customisation and configuration in a 'restabilised' usable organisational form, but that its usability will also require its consumption by a broad range of communities of interest whose understanding of and value they attribute to the technology will vary. In this sense, a 'stable' technology is one that has achieved 'some obduracy and users are able to develop the usability and utility of those systems to the extent that these can balance standardisation with multiple understandings and incorporations' (McLaughlin 1999: 217).

In all these discussions, it is argued that the obduracy and materiality of technology should be retained as an inextricable part of a recursive relationship within which resides the possibility for multiple interpretations and meanings (see Preece and Clarke, Chapter 3, this volume). Indeed, we would argue that these kinds of reformulations of the technology–organisation relationship move us towards a conceptualisation that maintains a notion of 'technology' as having independent effects in specific contingent circumstances and contexts but, at the same time, shows that these 'effects' are shaped over time by evolving constituencies and networks of socio-economic relationships and structures (what might be termed a socio-historical process with technical clout).

We would also suggest that the notion of the 'mutual shaping' of technology and organisation provides a useful broadening out into a more panoramic perspective of the conventional interests of organisational researchers. No longer does our attention focus on a need to identify and define a dividing line between the technical and social – preferably one that 'squeezes out' the technical from our analysis; but rather, concern shifts to the dynamic interpretative interplay surrounding the design, development and use of technology within and between organisations. As such, the non-linear process of technological development is open to a range of possible outcomes which may become stabilised/destabilised/restabilised in a variety of ways but where such outcomes are always contingent and ultimately transitory – indeed, the metaphor of technology crystallising in particular contingent circumstances has much resonance (Fleck 1993). In addition, we would suggest that the distinction between the production (design, adoption, implementation) and consumption (deployment and use) of technology, which in the past has not been widely deployed, deserves to be more fully exploited.

In sum, we may now regard the social shaping of technology as something which takes place both outside and inside the adopting organisation (see Dawson 2000: 51–56). Moreover, this involves both a process of technology production and technology consumption (again both inside and we would suggest outside the adopting organisation). Furthermore, the manner in which political processes of choice and negotiation – which we maintain still lie at the heart of this shaping process – are themselves shapers of and shaped by a broader network of socio-economic relationships and structures becomes easier to demonstrate. Indeed this may well, as Child (1997: 65–69) has noted, result in a more adequate understanding of the evolutionary need for organisations to learn generated by their environmental and technological circumstances, and the manner in which they do so or not – through the actual agency of organisational members. By the same token, the wide variations in actual organisational outcomes observed in workplace studies can now be better understood and explained in terms which go beyond the idiosyncrasies of the political systems and culture of the adopting organisation.

Conclusion

In this chapter we have attempted to throw some new light on the problem of understanding the technology–organisation relationship. We have shown how, from the supposed technological determinism of Woodward (1980) to the more voluntaristic and social constructivist accounts of Pinch and Bijker (1987), there remains dispute over the determining elements of technology. Early studies captured the materiality of technology and placed technology at the centre of analysis and explanation in studies on the impact of technology on organisations. As we have shown, during the 1980s and 1990s attention to choice and negotiation over the uptake and use of technology moved attention towards the social and political processes of change. Under the more social constructivist accounts, the materiality of technology became questioned and at times lost. Our contention here is that there is a need to bring the materiality of technology back into the equation in studies of technology at work. We have shown how organisational analysis has struggled to overcome the problem of technological determinism and the value of developing a broader mutual shaping perspective by importing insights from the sociology of technology. Appropriately adapted, such an approach offers the potential to move beyond the often implicitly determinist accounts of the technology–organisation relationship that much work (including our own earlier work) has been prone to produce. As such, a mutual shaping approach would take account of technology as both a material and socially constructed artefact and thereby avoid the difficulties evident in some of the social constructivist accounts which focus only on technology as represented in language and discourse.

The perspective we seek to encourage and develop further recognises the manner in which the characteristics and capabilities of technology can enable and constrain organisational transformation in particular circumstances, but also seeks to explore the interpretative flexibility of the 'technical' as it is configured and reconfigured in cycles of adoption, adaptation in use and redevelopment and application in other organisational settings. In other words, we are arguing for a mutual shaping perspective which recognises that while technology is designed and developed by a range of individuals and groups, these and others also engage in constructing meanings of technology within a changing social context. We stress the importance of temporal contextual influences and how, over time, there may arise a common understanding among different groups and individuals on what constitutes a particular technology. As such, there may be a form of stabilisation and closure in which the technology may appear obdurate. However, we contend that when technology is introduced into the workplace it is likely to be reconfigured by users and a number of configurations in use may arise in different organisational contexts and among different groups within the same organisation. This user shaping of the technology may in turn influence further design and developments in the ongoing mutual shaping of technology and organisation. In this sense, there is a duality of technology in which the social shaping of technology may at certain periods add to the obduracy of technology and there may be certain reconfigurations of technology in use which present a challenge to previous agreements on what constitutes technology. Such a perspective would take account of technology as both a 'hard place' – tangible, material entity – and as a socially constructed artefact ('cinema') – given meaning through language and other forms of representation in specific contexts.

In order to understand and study this dynamic movement between the social/symbolic and material, it is important not to be drawn away from the workplace or context of technology in use. Longitudinal ethnographic studies can capture those processes by which technology is stabilised and common interpretations emerge. Furthermore, we would contend that while there is no final end-point, there are certain periods in which technology is stabilised and common meanings are more broadly agreed across a wider range of stakeholders (suppliers, developers of technology, consumers). During these periods, the materiality of technology may serve to enable and constrain the social process of choice and negotiation in the uptake and use of technology. However, what is interesting about communication and information technologies is not only how they may be reconfigured and reinterpreted over time, but also how they may unlock previously stabilised technologies, such as the automobile, in enabling the possibility to reinterpret the meaning of the technology in question. For us, this raises important research questions about the materiality of technology which require further investigation. For example, what are the processes by which a certain technology becomes stabilised and obdurate? How does the materiality of technology shape common meanings on what a technology is? How does our understanding of a technology influence reconfiguration in use and how do post adoption developments influence our understanding of technology?

In attempting to answer these difficult questions, we need to move on from materiality found and materiality lost, to bringing technology back into a mutual shaping perspective which attempts to engage with these dynamic socio-historical processes.

Note

1 'Between Cinema and a Hard Place' is a work by video artist Gary Hill. The work consists of twenty-three television monitors devoid of their outer casings and arranged in a darkened room in lines. The screens show a sequence of apparently disconnected scenes and images attached to an equally disconnected soundtrack. Images transfer from screen to screen, flicker and blur, and screens go blank and switch on and off. The 'hard place' of the hardware contrasts with the immateriality and dislocated meanings of the 'cinematic' images. As the work progresses the link between the two becomes increasingly unclear (Tate Modern 2000). The work was exhibited at the Tate Modern, London, between 13 May and 4 December 2000 as part of an exhibition of the same name.

References

Badham, R. (1993) 'Systems, networks and configurations: inside the implementation process', *International Journal of Human Factors in Manufacturing* 3, 1: 3–13.

Badham, R. (1995) 'Managing socio-technical change: a configuration approach to technology implementation', in J. Benders, J. de Haan and D. Bennett (eds) *The Symbiosis of Work and Technology*, London: Taylor & Francis, pp. 77–94.

Baldry, C. (1988) *Computers Jobs and Skills: The Industrial Relations of Technological Change*, London: Plenum Press.

Barnatt, C. (1995) 'Office space, cyberspace and virtual organisation', *Journal of General Management* 20, 4: 78–92.

Bell, D. (1973) *The Coming of the Post-Industrial Society*, New York: Basic Books.

Bijker, W.E. and Law, J. (1992) 'Do technologies have trajectories?' in W.E. Bijker and J. Law (eds) *Shaping Technology/Building Society: Studies in Socio-technical Change*, Cambridge, MA: MIT Press, pp. 17–20.

Blosch, M. and Preece, D. (2000) 'Framing work through a social-technical ensemble: the case of Butler Co.' *Technology Analysis & Strategic Management* 12, 1: 91–102.

Braverman, H. (1974) *Labour and Monopoly Capital: The Degradation of Work in the Twentieth Century*, New York: Monthly Review Press.

Buchanan, D.A. and Boddy, D. (1983) *Organisations in the Computer Age: Technological Imperatives and Strategic Choice*, Aldershot: Gower.

Child, J. (1972) 'Organisation structure, environment and performance: the role of strategic choice', *Sociology* 6, 1: 1–22.

Child, J. (1997) 'Strategic choice in the analysis of action, structure, organizations and environment: retrospect and prospect', *Organization Studies* 18, 1: 43–76.

Clark, J., McLoughlin, I.P., Rose, H. and King, J. (1988) *The Process of Technological Change: New Technology and Social Choice in the Workplace*, Cambridge: Cambridge University Press.

Clausen, C. and Koch, C. (1999) 'The role of accasions and apaces in the transformation of information technologies', *Technology Analysis & Strategic Management* 11, 3: 463–482.

Clausen, C., Dawson, P. and Nielsen, K.T. (eds) (2000) 'Political processes in management, organization and the social shaping of technology', Special Edition, *Technology Analysis & Strategic Management* 12, 1: 1–143.

Davis, L.E. and Taylor, J.C. (1973) 'Technology, organization and job structure', in R. Dubin (ed.) *Handbook of Work, Organization and Society*, New York: Houghton-Mifflin, pp. 379–419.

Dawson, P. (1994) *Organizational Change: A Processual Approach*, London: Paul Chapman.

Dawson, P. (1996) *Technology and Quality: Change in the Workplace*, London: International Thomson Business Press.

Dawson, P. (2000) 'Technology, work restructuring and the orchestration of a rational narrative in the pursuit of "management objectives": the political process of plant-level change', *Technology Analysis & Strategic Management* 12, 1: 39–58.

Dawson, P., Clausen, C. and Nielsen, K.T. (2000) 'Political processes in management, organization and the social shaping of technology', *Technology Analysis & Strategic Management* 12, 1: 5–16.

Fleck, J. (1993) 'Configurations: crystallizing contingency', *International Journal of Human Factors in Manufacturing* 3, 1: 15–36.

Fleck, J. and Howells, J. (1997) 'Defining technology and the paradox of technological determinism'. Department of Management Studies, Working Paper, Uxbridge, Brunel University.

Forester, T. (ed.) (1980) *The Microelectronics Revolution*, Oxford: Blackwell.

Forester, T. (ed.) (1985) *The Information Technology Revolution*, Oxford: Blackwell.

Garrety, K. and Badham, R. (2000) 'The politics of socio-technical intervention: an interactionist view', *Technology Analysis & Strategic Management* 12, 1: 103–118.

Grint, K. and Woolgar, S. (1992) 'Computers, guns, and roses: what's social about being shot?', *Science, Technology and Human Values* 17, 3: 366–380.

Grint, K. and Woolgar, S. (1997) *The Machine at Work*, Cambridge: Polity Press.

Huczynski, A. and Buchanan, D.A. (1991) *Organisational Behaviour*, 2nd edn, London: Prentice Hall.

Kling, R. (1991a) 'Computerisation and social transformations', *Science, Technology and Human Values* 16, 3: 342–367.

Kling, R. (1991b) 'Reply to Woolgar and Grint: a preview', *Science, Technology and Human Values* 16, 3: 379–381.

Kling, R. (1992a) 'Audiences, narratives and human values', *Science, Technology and Human Values* 17, 3: 349–365.

Kling, R. (1992b) 'When gunfire shatters bone: reducing socio-technical systems to social relationships', *Science, Technology and Human Values* 17, 3: 381–385.

Knights, D. and Murray, F. (1994) *Managers Divided: Organisational Politics and Information Technology Management*, Chichester: Wiley.

Knights, D. and Wilmott, H. (eds) (1988) *New Technology and the Labour Process*, London: Macmillan.

Knights, D., Wilmott, H. and Collinson, D. (1985) *Job Redesign: Critical Perspectives on the Labour Process*, Aldershot: Gower.

Koch, C. (2000) 'The ventriloquist's dummy? The role of technology in political processes', *Technology Analysis & Strategic Management* 12, 1: 119–138.

Kumar, K. (1995) *From Post-Industrial to Post-Modern Society: New Theories of the Contemporary Work*, Oxford: Blackwell.

Latour, B. (1991) 'Technology as congealed social relations', in J. Law (ed.) *A Sociology of Monsters*, London: Routledge.

Lyon, D. (1994) *The Electronic Eye: The Rise Of The Surveillance Society*, Oxford: Polity Press.

MacKenzie, D. and Wajcman, J. (2000) *The Social Shaping of Technology: How the Refrigerator got its Hum*. Milton Keynes: Open University Press.

McLaughlin, J., Rosen, P., Skinner, D. and Webster, J. (1999) *Valuing Technology: Organisations, Culture and Change*, London: Routledge.

McLoughlin, I.P. (1997) 'Babies, bathwater, guns and roses', in I.P. McLoughlin and M. Harris (eds) *Innovation, Organisational Change and Technology*, London: ITB Press.

McLoughlin, I.P. (1999) *Creative Technological Change: The Shaping of Technology and Organisation*, London: Routledge.

McLoughlin, I.P., Badham, R. and Couchman, P. (2000) 'Technological change: sociotechnical configurations and frames', *Technology Analysis & Strategic Management* 12, 1: 17–38.

McLoughlin, I.P. and Clark, J. (1994) *Technological Change at Work*, 2nd edn, Buckingham: Open University Press.

Orlikowski, W.J. (1992) 'The duality of technology: rethinking the concept of technology in organisations', *Organisational Science* 3, 3: 398–427.

Parker, M. (1999) 'Capitalism, subjectivity and ethics: debating labour process analysis', *Organization Studies* 20: 25–45.

Pettigrew, A. (1973) *The Politics of Organizational Decision-Making*, London: Tavistock.

Pinch, T.J. and Bijker, W.B. (1987) 'The social construction of facts and artifacts: or how the sociology of science and the sociology of technology might benefit each other', in W.E. Bijker, T.P. Hughes and T.J. Pinch (eds) *The Social Construction of Technological Systems: New Directions in the Sociology of History and Technology*, Cambridge, MA: MIT Press, pp. 17–50.

Preece, D. (1995) *Organisations and Technical Change: Strategy, Objectives and Involvement*, London: Routledge.

Preece, D., McLoughlin, I. and Dawson, P. (eds) (2000) *Technology, Organizations and Innovation: Critical Perspectives on Business and Management, Volumes I, II, III and IV*, London: Routledge.

Roberts, K.H. and Grabowski, M. (1996) 'Organisations, technology and structuring', in S.R. Clegg, C. Hardy and W.R. Nord (eds) *Handbook of Organization Studies*, London: Sage, pp. 408–423.

Robey, D. (2000) 'Taking stock of the impact of information systems'. Paper presented at the Twenty-eighth Annual ASAC Conference/Fifth IFSAM World Congress, University of Quebec at Montreal, 8–11 July.

Russell, S. and Williams, R. (2000) 'Social shaping of technology: frameworks, findings and implications for policy', in R. Williams (ed.) *Concepts, Spaces and Tools: Recent Developments in Social Shaping Research*, COST A4: Focused Study on the Social Shaping of Technology COST-STY-98-4018. Edinburgh: Research Centre for Social Sciences/Technology Studies Unit, 27–106.

Salzman, H. and Rosenthal, S. (1994) *Software by Design: Shaping Technology and the Workplace*, New York: Oxford University Press.

Sayles, L.R. (1958) *The Behaviour of Industrial Workgroups*, New York: Wiley.

Scarbrough, H. and Corbett, J. (1992) *Technology and Organization. Power, Meaning and Design*, London: Routledge.

Sewell, G. (1998) 'The discipline of teams: the control of team-based industrial work through electronic and peer surveillance', *Administrative Science Quarterly* 43, 2: 397–428.

Sewell, G. and Wilkinson, B. (1992) 'Someone to watch over me: surveillance, discipline and the just-in-time labour process', *Sociology* 26, 2: 271–289.

Smith, C. and Thompson, P. (1998) 'Re-evaluating the labour process debate', *Economic and Industrial Democracy* 19, 4: 551–577.

Stewart, R. (1971) *How Computers Affect Management*, London: Macmillan.

Tate Modern (2000) *Between Cinema and a Hard Place*, Exhibition Catalogue.

Thomas, R.J. (1994) *What Machines Can't Do: Politics and Technology in the Industrial Enterprise*, Berkeley: University of California Press.

Thompson, P. (1989) *The Nature of Work*, 2nd edn, London: Macmillan.

Walsham, G. (1993) *Interpreting Information Systems*, Chichester: Wiley.

Wilkinson, B. (1983) *The Shop Floor Politics of New Technology*, London: Heinemann.

Willcocks, L., Currie, W. and Mason, D. (1996) *Information Systems at Work: People, Politics and Technology*, London: McGraw-Hill.

Williams, R. (1997) 'Universal solutions or local contingencies? Tensions and contradictions in the mutual shaping of technology and work organisation', in I.P. McLoughlin and M. Harris (eds) *Innovation, Organisational Change and Technology*, London: ITB Press, pp. 170–185.

Williams, R. (ed.) (2000) *Concepts, Spaces and Tools: Recent Developments in Social Shaping Research*, COST A4: Focused Study on the Social Shaping of Technology COST-STY-98-4018. Edinburgh: Research Centre for Social Sciences/Technology Studies Unit.

Woodward, J. (ed.) (1970) *Industrial Organisation: Behaviour and Control*, London: Oxford University Press.

Woodward, J. (1980) *Industrial Organisation: Theory and Practice*, 2nd edn, Oxford: Oxford University Press.

Woolgar, S. and Grint, K. (1991) 'Computers and the transformation of social analysis', *Science, Technology and Human Values* 16, 1: 368–378.

Zuboff, S. (1988) *In the Age of the Smart Machine*, Cambridge, MA: HBR Press.

3 Company intranets, technology and texts

David Preece and Ken Clarke

Introduction

Increasing attention has been devoted to the 'new wave' of information and communication technologies as they have become increasingly widespread in work and non-work organizations. One example is intranets, which may be viewed as the intra-organizational equivalent of the internet. Intranets represent a form of simultaneous technological and organizational change and engender a number of important questions with respect to the changing nature and experience of work in the early years of the twenty-first century, not least as to how these developments might be theorized.

This chapter is offered as a contribution to the ongoing debate about the nature and implications of information and communication technologies and the conceptualization of technology in its social and organizational setting. It also has the more specific aim of reporting some of our case study findings about the purposes for which, and the ways in which, intranet technology is introduced into organizations and contingencies of utilization and deployment.

Adopting, using and talking about intranets

An intranet is an architectural technology; that is, in itself it does not imply any specific uses. The latter are created and located within the enabling structure of the intranet, and thus a wide variety of applications are possible through an intranet platform (Newell *et al.* 2000, 2001). One common claim of progenitors is that it can be configured to act as a means of within-organization communication, enhancing collaboration and productivity (Wachter and Gupta 1997). It does this by acting as a publication medium which can be readily accessed and searched by users. It can also act as a medium for transactions across the organization via the page structure (drawing upon software), providing enhanced access and ease of use, and can be a medium of discussion and record. Thus, in summary terms, intranet technology has both wide integrative and distributive potential and lends itself to extensive creative and interpretive flexibility (or 'texts') as to its applications, uses, role and organizational impact.

Usually, an intranet configuration is specific to the particular organization within which it is created, developed and used, and within a given organization intranets are designed with quite different features (Newell *et al.* 2000). Thus, each configuration will be influenced by both the relevant social, economic and political contexts and the actions, orientations and concerns of the actors involved (Damsgaard and Scheepers 1999, 2000). It follows that the nature of the emergent configuration is, potentially at least, open to much contestation and interpretation and that instability is built in; that is, one should not expect to find a final, now fixed-for-ever configuration, but rather an ever-evolving one. In other words, there is little stabilization in the sense of a final state of closure, though there is considerable sedimentation as the technology becomes incorporated as the de facto means of operation.

Intranet technology may be classified as a form of 'groupware', along with similar ICTs such as Lotus Notes. As Ciborra (1996, 2000) has pointed out, the term 'groupware' implies the distinct, but linked, attributes of *group* as a social entity, and *ware* as the material artefact. Technology, it is argued, has both an artefact or material form, and is also a social construct, whereby its design, development, creation and use is the outcome of human agency. This view of technology has been widely explored and commented upon in the literature, most commonly through social construction of technology (SCOT), social shaping of technology (SST) perspectives, and actor-network analyses (see Chapters 1 and 2, this volume). The main challenge to this view has come from authors who take up a technology as text and metaphor (TTM) position (see Grint and Woolgar 1997 and Chapter 2, this volume), wherein the technological/social configuration is never stabilized, always being open to different and new interpretations or 'texts'. In this chapter we theorize intranets through the concept of a socio-technical ensemble (influenced strongly by SCOT/SST perspectives; see e.g. Bijker 1995), where the 'component parts and their composition are shot through with and held together by social relations among people, as much as by more physical ties such as screws, bolts or electrons' (McLaughlin *et al.* 1999: 6; see also Blosch and Preece 2000).

The socio-technical ensemble, then, is socially created and shaped by the ideas, interests, objectives of and interactions between designers, technologists, engineers, managers, users and other actors. This social shaping takes place both as the technology generically develops in an inter-organizational context, and as it is specifically configured within particular organizational contexts. This configurational process continues as the technology is introduced, implemented and put into everyday use (Fleck 1994; McLaughlin *et al.* 1999). Certain actors and groups have more opportunity than others to shape the technology–organization configuration during implementation and deployment. It is therefore important to know who has this opportunity, what their aims are, what sources of influence they have, and what means they deploy in order to shape the ensemble. The TTM perspective offers a route into exploring this: since the technology cannot speak for itself, one is compelled to explore what the people say who speak for it, who these people are, and what their background is. Among other considerations, this is central to the understanding of how certain actors, not least managers and IT specialists, attempt to legitimize technological/organizational change.

But how and where are the differences implied in the above argument played out? Figure 3.1 summarizes the key elements of the intranet socio-technical ensemble which emerged from our case study research, and the arena in which contests over meaning, interpretation, significance and action take place. In developing this framework we have drawn on a number of strands in the literature and made some preliminary attempts at identifying some of the important linkages between the various aspects/elements. The six main elements are: (1) technological malleability; (2) initial acceptability; (3) configuring form and function; (4) usability; (5) boundaries, and (6) sedimentation. Briefly taking each in turn:

1 How much flexibility is there for shaping and reshaping the technology, and what are the main dimensions and constraints to that process? For example, is the technology seen as generic, with little possibility for reshaping, or configurational, that is, with much more flexibility built in?
2 What is the degree of initial acceptability of a new technology, and what influences this? For example, to what extent do initial perceptions see the ensemble as disruptive of present practice, as against allowing for a continuity of practice?
3 What form does the ensemble take and what utility does it provide for the organization in general and for the people who work in it? To what

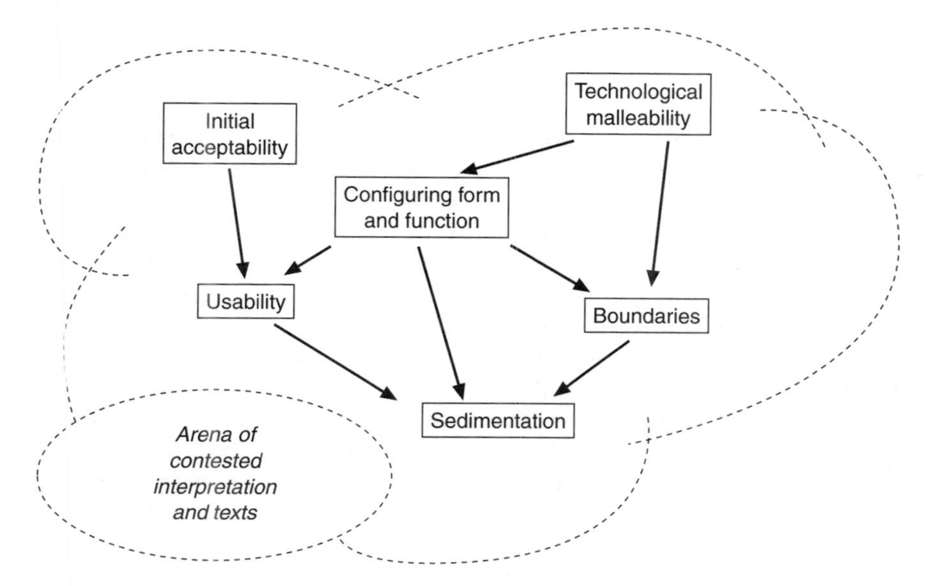

Figure 3.1 The intranet socio-technical ensemble

extent (if at all), and in what senses, does the configuration of the technology absorb and reflect the organization, mirroring and/or changing structure, processes and individual tasks?

4 How literate and confident are people in the use of the technology? How can we begin to understand variations in the levels of confidence and literacy?
5 Are the ensemble boundaries perceived as being open and transparent, or subject to forms of control and restricted access? Who is allowed to do what with the technology?
6 To what extent and, if so, in what respects does the intranet become an aspect of everyday working practices within the organization? If this does not occur, at least in some parts of the organization, why is this?

It is only possible in the space available to us to provide a flavour of the empirical data we have collected under each of these headings (which, of course, are used for analytical and presentational purposes, and are more provisional than this format might be seen to imply: for one thing, there is a great deal of overlap and iteration between the elements). Below (again primarily with a view to confining our discussion within word-limit bounds) we have conflated the broad scope of our framework into three main time phases or stages: intranet adoption (technological malleability, and initial acceptability); introduction and implementation (configuring form and function, and usability); and boundaries/sedimentation (for a much fuller outline and discussion of the framework see Clarke and Preece 1998; Preece and Clarke 1998, 2000). We draw here upon just one of the longitudinal company case studies which we have undertaken.

The case study

The primary data were gathered via interviews, documentary analysis and non-participant observation from a high-tech company based on the south coast of England (here referred to as 'Grangers'). Until shortly before our study Grangers had been a division of IBM (UK), and was created through a management buyout, continuing to manufacture computer hardware and software.

The data were collected through two series of interviews, the first conducted when the intranet first went live throughout the organization between February and early July 1997 and the second around six months later. Interviewees were taken from horizontal and vertical slices of the organization and were seen individually for between thirty minutes and two and a half hours, forty interviews being held in total. The interviews were conducted in a semi-formal manner, guided by just a few subheadings/areas for exploration (which were covered with all interviewees), with the interviewer always being happy to let the discussion develop along the lines chosen by the interviewee. The great majority of the interviews were, with the permission of the interviewee, tape-recorded, and subsequently transcribed verbatim. The interview transcriptions were read, coded, discussed, reread and re-coded by the authors a number of times. We also had access to some company documentation and undertook a limited amount of non-participant observation.

In the interview extracts below, after each quotation the interviewee's role is indicated, along with either a '1' or a '2': '1' indicates that the interview was conducted during the first round of interviews between February and July 1997, and '2' indicates a second round interview between January and March 1998. This provides an indication of whether the intranet was in the process of adoption or implementation at the time of the interview, or whether it had been in existence for a minimum of six months at the time of the interview (4 July 1997 being 'Independence Day', when the switch-over was made from the previous ICT configuration of the company, called the 'VM' or 'PROFS' system). 'M' = Manager, 'D' = Director.

Phase 1: Intranet adoption

What the intranet offers, both initially and as it develops, has much to do with (1) the way it has been set up by people and (2) how people configure its subsequent use. This design flexibility in the overall architecture influences many later parameters in the evolution and use of the intranet, such that perceptions of this freedom of design become significant. Within this architecture there was seen to be much room for choice:

> They try to keep it operating such that the building fabric is provided by IS and what you do inside the house is pretty much up to you, you can put your own carpets and curtains, but you are bounded by where the walls are.
>
> (IS and Quality Line M, 2)

At the same time, a particular group of people, in particular IS specialists, took the leading role in the configuration of the intranet in the early days and in designing the architecture which still constrains what is possible. While our study was in progress the firm moved to a business-centre organization structure and the intranet technology facilitated the decentralization of the IS function to the new divisions which were created. This would apparently have been more difficult to do, if not impossible, with the previous IT:

> First time that Grangers has swung away from being totally centralized to decentralized. One of the areas that got decentralized into the divisions was IS . . . you probably wouldn't have done this last year with the systems we were on'.
>
> (IS and Quality General M, 2)

> . . . you can easily migrate one business unit on to one server, taking those people and their procedures from this environment . . . and move them very rapidly on to a server to be a self-functioning organization.
>
> (Network Systems Line M, 2)

What, however, about the sharing of data and information across the organization following divisionalization: what role might the intranet play here? In the view of the Business Services Director it could help, but only insofar as it is set up to do so for specific groups of people:

> So I think the intranet will hopefully encourage sharing of data and using common files across large groups . . . so I will have an area on the intranet that myself and my direct reporting managers can access and nobody else can access, and we will share information in there, then we have wider groups, larger departments and functions that share common space on the intranet.
>
> (Business Services D, 1)

The intranet adoption process was triggered by the perception that the pre-existing technology had become outmoded in terms of its capabilities. Pressure built up, for example, from parts of the organization (not least recent graduates) to adopt more up-to-date forms of ICT. Among the main adoption advocates and drivers we found not only a representation of the intranet as radical in terms of its social and working practices implications, but also a recognition that the latter will emerge not just 'magically' from the technology per se, but rather from the socio-technical ensemble of which it forms a part:

> A little while back, when I presented what we are trying to do with the intranet, in the auditorium to the employees, I said it would be a cultural shock wave that would go through the company and people looked at me a little surprised, and I think they are beginning to realize now what was meant at the time.
>
> (Business Services D, 1)

> I can choose to work completely differently, and use this as a tool, but if I do work completely differently, won't come from the fact the tool exists, it will be somebody coming and knocking on my door and saying 'for god's sake, don't you realize you could be driving to France? . . .' I'm a great believer in catastrophe theory-type of models, and I think suddenly somebody will do something a bit different, everybody else will think 'oh, that's good', and suddenly it will take off.
>
> (IS and Quality Line M, 2)

Thus, in terms of what actually happens, much will depend upon the particular orientation adopted by staff towards the ensemble:

> Hopefully some people will seize the opportunity, they will see what's happening around them and they will think 'Well, this is a better way of operating', you know, the window of opportunity, the glimmer of light is there. . . . If they don't, we will fail and we will slip back into the old ways of kind of mass communication through the email system, etc.
>
> (Business Services D, 1)

Phase 2: Introduction and implementation

The IS and Quality General Manager, who played a leading role in the adoption and introduction of the intranet, recognized that while the technology architecture was a key part of the socio-technical ensemble, it would be the response and orientation of staff which would 'make or break' the change:

> I think the architecture is right, the equipment is right . . . but what will make this not work is the people will get in the way, or you will believe the people will get in the way . . . [but] put the two together and this is a winner all the way.
>
> (IS and Quality General M, 2)

One of senior management's key objectives in introducing the intranet into this recently formed company was to use it to help shape the new corporate identity, which they wanted to be seen as distinct by employees, customers and suppliers alike from the (remaining) legacy of its IBM parentage:

> So I think or I hope one of the things we will do with intranet is break some of that culture. It will cause people to stop and think and I think that's one of the things on the cultural side, in terms of whether it's been a success, in terms of the Grangers' identity, it depends on how people use it.
>
> (Business Services D, 1)

It was recognized that it was crucial to win the support of the CEO for the adoption of the intranet:

> The person who wants to distinguish this, to create a new identity away from IBM . . . is the chief executive of the company, and if he's the only one that says it looks and feels different, that's good enough for me, but I just know that our customers will no longer see those green screens and think IBM.
>
> (IS and Quality General M, 1)

For a number of the people whom we interviewed, the company intranet had been effective in helping to create this new identity: 'I think it's one of the most important changes in turning Grangers from an essentially IBM-like company into a small more fast-moving, more vibrant, more technology minded company' (IS and Quality General Manager, 1). Moreover, the new identity could be the precursor to further change:

> I think having things like our own home pages, and it just feels like our system and not the system that many of us were used to for many, many years, it makes people think slightly differently, and I think whenever you have a change event, and you will know this, it allows you to have, you can use it as a way of putting in other changes like . . . so I think yes the technology is very powerful.
>
> (Personnel Director, 1)

Some of the Directors wanted to use the intranet to encourage managers and staff to reflect upon their jobs and how they performed them, looking for new ways of working:

> I think that will come as we gain confidence and experience with it, you know, IS are helping us to jump start that process, I think, and then very quickly other people will have their own thoughts and ideas, and will start to shape the system, but some may be able to do that today because they are already familiar and knowledgeable with this kind of technology.
>
> (Network Services General M, 1)

Senior management also intended that the intranet would help to get staff directly communicating and interacting with each other more often than was the case with the old technology:

> I don't want to overplay this, I am not trying to say we shouldn't be using an email or an intranet solution, what we are trying to do is moderate people's behaviour on it so they really use it where it's effective and where it's less effective that they get back into actually you know, using the phone and just walking out and talking to people.
>
> (Business Services D, 1)

As far as top-down communication was concerned, the common perception was that senior management were using the intranet for corporate communications purposes:

> They're pretty good with the media stuff that management want to communicate down through . . . whereas today you switch on and it's there so there is an immediate reaction so you go on to it and click on to see what they are on about, you can do that in a couple of minutes. In terms of finding the bigger issues, and some of the ways around it I don't think that's happened.
>
> (Procurement Line M, 2)

With respect to what the intranet is used *for*, staff were expected to be more proactive in actively seeking out the information they wanted. The hope was that the new technology would act as a catalyst for people rethinking their roles and working practices through a process of exploration. Of course, the nature and extent of take-up varied across the organization:

> Now a lot of people will never actually press the first button, and so that's why I think you will have groups of people that find it very exciting, will make radical changes in looking at the way other people do things and putting their processes up there and sharing information, [but] whether that will evolve through the whole organization, I am not sure, because there's no particular encouragement for it at the moment.
>
> (Business Planning M, 2)

Some senior managers believed that the intranet would enable staff to get a more rounded view of the organization:

> My experience is that people will look, people are anxious to learn, the vast majority to me that's helpful because all that does is encourage people to think differently, be aware of other parties in the organization, hopefully break down some of the divides of 'oh well, its their problem, not mine', a chance to see where people fit into the overall process of the business, right? It can only help in that regard.
>
> (Manufacturing D, 1)

The intranet was seen as enabling new ad hoc structures to evolve through encouraging staff to explore web page data and reflect upon what they find:

> Anybody who is working on a particular project can do a quick scan into the web page – 'Who can help me on the subject?', and pick up the phone. So they have created their own little network.
>
> (Manufacturing D, 2)

On occasion, the interviewees seemed to say or imply that the organization itself or a part of it was now to be found *on* the intranet (not 'merely' represented there):

> It's very accessible, it's much more friendly to use, and its more interlinked. We've got the whole quality management system in there now, whereas previously it was a quality management system by name only; now we use it for documentation changes, for our organizational structure, we've got our calibration records on there . . . it's now a real quality management system.
>
> (Procurement M, 2)

With respect to the implementation of the intranet, the Board made the IS Manager responsible for rolling it out across the company. While the time-scales were tight and no 'hostages' were to be taken along the way, most people whom we interviewed thought that this was necessary and justified:

> I think the Director of IS and his group . . . drove that very vigorously with very tight timetables to make it work, and bluntly didn't take any excuses along the way. It was going to go in, you were going to be off the system by this date. . . . And I think that was extremely important because I think it could have been just . . . 'oh well, you can stay on it'.
>
> (Finance and Contracts General M, 2)

Notwithstanding this, IS acted quite subtly and strategically in working to get some early 'wins' which they knew would then be spread around through the organization's grapevine, and could be used as illustrations of what was possible and the benefits of the intranet:

> I wanted some people on the side, some small wins very quickly and picked one or two to be honest . . . the secretaries are our life-blood, and what do they do most of their days? Well, you know, documents and diary management, so I said 'Look, what do you think of this?' And within two minutes they had taken this off me and were saying 'How can you show this?', and so we slowly got two or three of them on our side and we educated the secretaries to try and use it.
>
> (IS and Quality General M, 1)

Shortly after this, IS introduced user representatives (who were selected by IS for their anticipated enthusiasm for the technology) from the various business units to act as disciples and carry the message deeper into the organization, and to act as sources of local expertise and two-way communication between the users and the IS specialists. As might be anticipated, the representatives interpreted their role in a variety of ways:

> Some of them are obviously better than others and are selling it, I mean there are a couple of areas which I think in a year's time, if it goes the way that business unit rep wants it to go, it will knock our socks off. And there's a couple of areas where we are going to have to go in and kick people, and they will do the basic minimum as grudgingly as possible. But that's life.
>
> (Quality Management Systems (QMS) M, 1)

As pockets of expertise built up around the organization, IS staff began to adopt a lower profile with regard to intranet introduction. The business units now became the main shapers and developers of the technological ensemble, albeit that, given the organizational restructuring (including to the IS department), all the main units now had an IS specialist working directly for them:

> IS has become less and less strategic in its own right and more of a commodity because of the technology that is available. You can generate complex networks almost on a commodity basis, the real driving seat now will be because of the business units in taking the technology to where its customers demand. . . . The strategic management of IT will vary in strength and intensity depending on the needs of the particular business unit, so some business units wouldn't need to do anything – the infrastructure is there, it ticks along.
>
> (Test Systems Line M, 2)

For some managers and staff in the business units this was not anticipated to be an easy transition, or one that they would necessarily choose to make in a personal sense. However, they may find it difficult to desist:

> There are a couple of areas who are not doing enough. When all the other areas have got theirs in, it's going to be blatantly obvious to anybody on the site who goes into it and finds a blank page . . . that someone has failed.
>
> (QMS M, 1)

Phase 3: Boundaries and sedimentation

Some staff said that the intranet had created a greater degree of transparency within the organization about what was happening, yet at the same time they recognized that boundaries had been placed around what was revealed; occasionally they offered a view which came close to collapsing the organization into the technology (and vice versa), and thus alluding to a virtual organization (see Chapter 1, this volume):

> What it will actually do, it's the first time that we will ever have the running of the company so publicly available to the people in the company to understand what's going on, because they will be able to get to it, if they want to get to information, and see what the company is like and what the company does and how it is structured . . . it's actually going to be there for them for the first time ever. . . . I mean there's obviously going to be some things that they're going to hide, I mean we are not going to have the finance things and the negotiations with companies, that's never going to be up.
>
> (QMS M, 1)

However, even this partial transparency left some managers feeling uneasy, and concerned that there is a lack of control over and consistency in the information presented, to say nothing of the always-open possibility of different interpretations being put on the same set of data:

> It's a two-way sword in some respects. I mean, information is much more readily available, it cuts down on costs, it makes people more productive; however, it's open to interpretation, because you can be very clear on what you are putting up there, so if somebody interprets the information the wrong way, without asking the questions, then you are in trouble. Also, it's open to the opportunity for people to publish potentially conflicting information if you are not careful, because you have got one group publishing their own set of results and the other group putting a different gloss on it, publishing their own set of results. Which is right?
>
> (Finance Line M, 2)

Others drew a distinction between reading and sharing information/data through the intranet, and doing real productive work:

> Part of our culture, our heritage is about efficiency and getting the most out of our people and talking. You know, 'People are important' type of stuff, real corporatey-type messages. It's salt of the earth, exactly what we need, but, frankly, we don't know what our people are doing, we are talking about efficiency. Bloody hell, you know, if anybody has got an hour a day to spend on this, then they have got an hour a day to do something quite productive. Scary. And we had our fair share in IS of, ah, miscreants, because they understand the technology as much as anything.
>
> (IS and Quality General M, 2)

So, what approach did senior management take to these matters? According to the Manufacturing Director, they adopted a 'stand-off' approach to begin with in order to allow staff the room to experiment, and then relied on middle managers to identify who was using the facility in a 'productive/imaginative' way and who was not, or, perhaps even worse, not using the technology at all:

> Again, we consciously decided we were going to allow people to learn, discover, explore in the short term, then rely on management to start pulling in the reins and recognizing where all these areas of potential lie within the system . . . that we nurture these people and we identify them quickly and that's why we are spending time educating the managers to recognize who the people are who are pushing the boundaries . . . and those who are not logging on, not even turning on the system and interacting with it, and there will be two extremes.
>
> (Manufacturing D, 1)

The objective of local control was linked to the careful selection of the business reps discussed above:

> We picked the business unit reps . . . they are mostly senior professionals with a smattering of managers in among them . . . they will control the people who put the documents up on the screen.
>
> (QMS M, 1)

The information put on the business unit web pages was seen by some senior managers as an indication of what people thought was important:

> Yes, I do look at some of the home pages and things that are designed by people, yes I do have a look. I have a vested interest in looking 'cos I like to see what's on there, and what's going on with respect to the development of some of the home pages, so yes I do . . . it gives you a better idea of what people think is important.
>
> (Personnel Director, 2)

Another justification put forward for having some controls/boundaries put around the intranet was to do with the establishment of commonalities in website construction and presentation, essentially an argument about efficiency:

> I think we have got to be careful we don't shape it to anarchy. I think that in order that we communicate efficiently and effectively we have to perhaps put some boundaries around, or some process even, around the way in which we want to create websites or documents, or what have you, because there is a benefit for the uniformity of approach when you are talking about communicating within a large organization.
>
> (Test Systems General M, 2)

Others drew upon one of the key reasons put forward by senior management and IS for introducing the intranet – that is, to encourage a climate of experimentation and entrepreneurialism – as an argument as to why little control/surveillance should be imposed:

> IS are in control of being able to turn off illegitimate use of the system if they wanted to, but I don't think they will, I mean you defeat the object to some extent . . . of an open access system, if you start trying to control it. So, apart from not having anything illegal or immoral or other such literature on it, it becomes a shame if you start limiting people's ability.
>
> (Business Planning Officer, 1)

Sometimes control was seen more in terms of the direct, personal control staff were able to exercise over intranet usage:

> I think you feel in control until you get a glitch, and then you feel totally out of control when you are sat there, your thoughts are flowing, you are working hard, and then suddenly it just ceases in front of you and nothing will happen, and rage sets in. Until then you feel totally in control: 'Super, I'm driving it, it's doing what I want it to do.' Then it occasionally gets its own back – it's like a living being.
>
> (New Business Staff Engineer, 2)

Whatever senior managers and IS specialists may say with respect to the question of intranet control, some staff harboured suspicions about what they are actually doing:

> My director might surf around the structure and ask simple questions, that you think 'Oh god, what's he found, what's he uncovered this time?' He can do that to his heart's content, yeh, he's very good at it, ha ha – unfortunately.
>
> (Testers Line M, 2)

This person is right to be suspicious, for IS does have a surveillance facility vis-à-vis the intranet:

> We had the ability centrally to get past people's passwords and get into their own account, but IS clearly had the ability to do that and we have retained that ability, probably it's a little stronger on the intranet. . . . So there's a kind of 'Big brother is watching you' from IS. . . . What we don't want to do in the early days is to abuse that, but the concern is you are going to get a proliferation of non-business or peripheral activity. . . . It is not intended to be punitive or penalize people for misuse of the system . . . but clearly that is something that can be done.
>
> (Business Services D, 1)

Many employees have extensive expertise in ICT and hence, whatever checks are built in, if they are determined they can probably get around them:

> I thought we had a fire wall that would stop a lot of this stuff coming in, but I'm told that you only need to be fractionally computer literate to step around it, and that's the problem we have if it's the generation that have been brought up on these machines, they just hack their way through and around a lot of this stuff. Why shouldn't they?
>
> (Test Systems General M, 2)

What could senior management do about this, assuming they wanted to? Even a management audit might be problematic:

> Where do you strike the balance? I don't know. Is it through management audits, which we did on the old system, because there was the possibility of people getting files and sending them around? So, I think a management audit, and say 'Show me your list of bookmarks'. . . . [However] It's pretty obvious if somebody has put a hammer through a coffee machine, it's not so obvious if somebody has got ten thousand bookmarks.
>
> (Test Systems General M, 2)

By the time of our second round of fieldwork there were indications that, for some people at least, the intranet was becoming sedimented into their everyday working practices; hence, at least pro tem, some degree of stabilization had occurred vis-à-vis the intranet ensemble for these members of staff. For example:

> When I first got involved I thought this is not going to work very well, but very, very quickly you got into the swing of it, and it was obvious that it wasn't going to be terribly difficult to get used to . . . it has just come in and we are using it just like it was there forever.
>
> (Procurement Line M, 2)

> It feels as though it has been with us a lot longer than it has. It's interesting if you think back about the old legacy systems and the IBM VM system – it seems a long time ago. I think there is a level of comfort that has come with the intranet.
>
> (Business Services D, 2)

Discussion

The choice of technology, its adoption, introduction, implementation and use within the organization, all takes place within a social, economic and political arena of contested interpretation, texts and meanings; in other words, in a socio-technical ensemble which is inherently unstable, and thus where any stabilization is at best temporary but in any event likely to be only partial, i.e. not organization-wide. The texts which are 'attached' to the technology are constructed by certain people, with their particular bundles of experience, skills and knowledge, and with certain objectives in mind (organizational and/or personal), and the ensemble continues to evolve accordingly over time. This view has much in common with a 'technology as text and metaphor' perspective; yet, through the case study, we also see why it is necessary to preserve the facticity of the technology per se, and hence to talk about 'technology *and* texts', rather than 'technology *as* text(s)'. Thus our preference for theorizing company intranets as 'socio-technical ensembles', but seeing them as inherently unstable and subject to a variety of sometimes contested interpretations, and with a range of texts attached to them, which themselves change over time as a result of a mutual shaping process.

At Grangers, customers were insisting that electronic links be established with them, while some staff, especially the new graduate entrants, were pushing for it both internally and externally. At the same time, senior management wanted to demonstrate that the company had entered a 'brave new world' quite separate from IBM, and the severing of the old technology's umbilical link into IBM was one way in which this could be done – but this meant it then needed an ICT of its own (and a not too expensive one at that). The IS department was charged with the responsibility of investigating the possibilities and alternatives, not least because it had the relevant locus of expertise and would be designing the basic 'shell' or 'engineering system' (Clark *et al.* 1988) of the new ICT configuration, within the parameters of which users would subsequently have to work. Thus this handed a lot of influence over the identification, selection and design of the new technology and the adoption process itself to the IS department. Over time, as we move to implementation and day-to-day utilization, this influence over the intranet configuration diffused somewhat, in the main as a result of the organizational restructuring and the widening of the user network. The extent and quality of the latter was in part a function of users' confidence in and knowledge of the technology. This had a lot to do not only with their particular job, but also with their previous experience of this and other related forms of ICT both inside and outside the organization – the familial situation appeared to be especially influential here (for example, wanting to teach/help one's children or even learning from children in some cases!). The IS Manager devoted time at the early stage of implementation to demonstrating the benefits of the new system to selected secretaries, in the expectation that they would readily see how the intranet could benefit their work. Similarly, the user representatives were selected by IS staff as, *inter alia*, likely enthusiasts. It was assumed – correctly, as it turned out – that both they and the secretaries would 'carry the (positive) message deep into the organization' (Pettigrew and Whipp 1991).

Over time pockets of expertise built up around the organization, and some of the business units came to be the main 'drivers and shapers' of the intranet configuration. This was helped by the fact that the majority of IS staff were now working directly for the business units. At both the business unit and individual levels, it appears that one important reason why some people were using the intranet was that they perceived they needed to be seen to be doing so and/or that this would reflect positively upon management's perception of them. Group and peer pressures were probably operating, at both localized (within business units) and wider levels (across the organization and in comparison with other business units).

The intranet was represented initially (Jackson 1997) by senior staff (not least IS specialists) as being a radical innovation and as symbolic of 'how things have changed' and 'how they need to change' from when the company was owned by IBM (as one IS specialist said to us: 'customers will no longer see those green screens and think IBM'). This establishment of a new corporate identity was argued to be the main legitimation for the 'top-down' imposition strategy of introduction, to tight time schedules. None the less, some people experienced the intranet ensemble as an incremental change; they often commented that they obtained at least as much functionality from the old technology. Many of the staff whom we interviewed as well as ourselves could only really make sense of this company intranet *through* the job and working practices changes which went hand-in-hand with its implementation (as one person put it, 'it was a very cultural thing'). It was seen by many as having little relevance for some staff (for example, those with Taylorized, rule-bound jobs), as against others who were *expected* to be innovative. However, in practice, this turned on the question of whether such people *chose* to be innovative in their use of the technology, and this depended in part upon their degree of technological literacy.

The response of some people was to choose (as compared to the situation with the previous ICT) non-ICT facilitated means of communication, for example by spending more time talking face-to-face with colleagues. This may turn out to have been a short-term effect while people were adapting. However, senior managers frequently viewed this as a desirable state of affairs, insofar as email traffic is reduced and the talk is about work-related matters.

By the time of our second round of data collection, when the intranet had been in use for over six months, a conflict had emerged between the wish of some people for standard operating procedures and templates (for example, in order to facilitate the auditing process) and the view of some of these same people and also of others that precisely one of the key organizational benefits realizable via the intranet was the spur it was seen to be able to provide to innovative activity. Thus, for example, it was seen as encouraging the sharing of data and information across the organization – but then there were limits as to how far management wanted to go in this direction. Even at this relatively early stage of 'innofusion' (Fleck 1994), a few people were representing the intranet to us in a way which effectively collapsed it into the organization and vice versa (for example, 'It's the first time we will ever have the running of the company so publicly available'). Others identified certain dangers here, such as the fact that the information on the web pages had not been validated and/or was being 'misinterpreted' by users. We also got some 'hints' that there may be some 'self-serving' purposes behind what people were saying through their web pages. It was recognized that having the facility to put up information in this way for 'all' the organization to see did not necessarily mean that the company was more 'open': key questions were 'What information is put up?' (and, contrariwise, what is not put up), and 'What is the quality of that information?'

Our data indicate that overall, the intranet undermines attempts senior management might make to relay and mediate in both information communication and its attached meaning; many of the comments we received from senior management expressed their disquiet about these matters. Increasingly, the transparency built into the intranet ensemble, allowing employees to select from the various domains of information *directly* available to them, means that in a sense they are constructing through usage their own view of the organization and its members. The intranet offers more possibilities than the old technology about what information is made available and how that information is mediated. At the same time, as we saw, there is much 'sensitive' information which senior management does not put on the intranet, and employees recognize this; equally intranet usage is being monitored by IS, both in terms of what is put up and by whom (and who is not putting anything up), and in terms of what members of staff are accessing through the intranet and internet.

McLaughlin *et al.* (1999) comment that 'there is considerable pressure by management to ensure that systems are "de-localized", that is, deployed in standardized ways, precisely because this is where they see the value of MIS technologies lying'. This is not always the case, however: yes, there are pressures in this direction from some managers, but they and other managers might also want to simultaneously encourage innovatory activity at the job level. There is an obvious tension here. The Manufacturing Director, for example, told us that business unit managers had been 'encouraged' to monitor who in their unit was using the intranet, and for what purposes (including innovative ones). Of course, senior management did not want the intranet/internet to be 'abused' by, for example, 'glossy pictures' being pulled down. Yet there was also a recognition that they could well defeat a key objective of intranet introduction, namely encouraging innovative behaviour related to business objectives and methods, if they regularly issued edicts about its use and either directly or indirectly (via IS) adopted a high-profile surveillance position. However, whatever they and IS staff *say* about intranet regulation, it would appear that their staff will always harbour suspicions about what is going on – and rightly so, as we have seen (IS do have a surveillance capability, which they use). Having said that, there are some very computer-literate people in this organization, who are probably able to circumvent whatever controls or monitors are put in place *if they so choose.* Indeed, the extent to which this was happening (and, for obvious reasons, we do not have data on this) could be taken as an indication that the intranet socio-technical ensemble was sedimenting into the organization.

References

Bijker, W. (1995) *Of Bicycles, Bakelites and Bulbs, Towards a Theory of Socio-Technical Change*, MIT Press, Cambridge, MA.

Blosch, M. and Preece, D. (2000) 'Framing work through a socio-technical ensemble: the case of Butler Co', *Technology Analysis & Strategic Management* 12, 1: 91–102.

Ciborra, C. (ed.) (1996) *Groupware and Teamwork: Invisible Aid or Technical Hindrance*, Wiley, Chichester.

Ciborra, C. (ed.) (2000) *From Control to Drift: The Dynamics of Corporate Information Infrastructures*, Oxford University Press, Oxford.

Clark, J., McLoughlin, I., Rose, H. and King, J. (1988) *The Process of Technological Change: New Technology and Social Choice in the Workplace*, Cambridge University Press, Cambridge.

Clarke, K. and Preece, D. (1998) 'The company intranet: a case study in evolution and control'. Conference: Constructing Tomorrow: Technology Strategies for the New Millennium, University of the West of England Business School, September.

Damsgaard, J. and Scheepers, R. (1999) 'Power, influence and intranet implementation: a safari of South African organisations'. *Information Technology and People,* 12: 333–358.

Damsgaard, J. and Scheepers, R. (2000) 'Managing the crises in intranet implementation: a stage model'. *Information Systems Journal* 10: 131–149.

Fleck, J. (1994) 'Continuous evolution: corporate configurations of information technology', in R. Mansell (ed.) *The Management of Information and Communication Technologies: Emerging Patterns of Control,* ASLIB, London.

Grint, K. and Woolgar, S. (1997) *The Machine at Work: Technology, Work and Organization*, Polity Press, Cambridge.

Jackson, P. (1997) 'Information systems as metaphor: innovation and the 3 Rs of representation', in I. McLoughlin and M. Harris (eds) *Innovation, Organisational Change and Technology*, ITB Press, London.

Pettigrew, A. and Whipp, R. (1991) *Managing Change for Competitive Success*, Blackwell, Oxford.

McLaughlin, J., Rosen, P., Skinner, D. and Webster, A. (1999) *Valuing Technology: Organisations, Culture and Change*, Routledge, London.

Newell, S., Scarborough, H., Swan, J. and Hislop, D. (2000) 'Intranets and knowledge management: de-centred technologies and the limits of technological discourse', in C. Prichard, R. Hull, M. Chuner and H. Willmott (eds) *Managing Knowledge: Critical Investigations of Work and Learning*, Macmillan, London.

Newell, S., Scarborough, H. and Swan, J. (2001) 'From global knowledge management to internal electronic fences: contradictory outcomes of intranet development'. *British Journal of Management* 12, 2: 97–111.

Preece, D. and Clarke, K. (1998) 'Representing and using an intranet: a case study'. Eighth International Forum on Technology Management, Grenoble, November.

Preece, D. and Clarke, K. (2000) 'Company intranets: technology and texts'. Sixteenth EGOS Colloquium, Organizational Praxis, Helsinki School of Economics and Business Administration, Helsinki, July.

Wachter, R. and Gupta, J. (1997) 'The establishment and management of corporate intranets'. *International Journal of Information Management* 17: 393–404.

4 ERP software packages

Between mass-production communities and intra-organizational political processes

Christian Koch

Introduction

The market for Enterprise Resource Planning systems (ERPs) has significantly changed over the past decade. At least in Denmark, manufacturing enterprises, which previously were used to close 'partner-like' collaboration with their IT supplier, now face mass-producing software houses and their allies among consulting companies, hardware suppliers, customers and so on. These communities challenge the management of technology at the organizational level, since configuring generic packages, and shaping them to meet the companies' diverse needs, is a major task. This constitutes political processes of alliance building, choice and compromise. This chapter's aim is to analyse how the communities impact on company implementation. In doing this, it combines two long-term processual fieldwork studies of political processes at plant level, twenty-five shorter case studies of implementation and use of ERP, and a macro-oriented study of the three ERP communities around the three systems *SAP R/3*, *BaaN IV* and *Navision XAL*.

There are marked differences between the three ERP communities. The large groups implementing *SAP R/3* have begun virtualization of their organization using the system to enable geographically spread units to operate in consorted action. Some centralization of organizational functions such as purchasing, finance and IT accompany this. On the other hand, this community is not characterized by delegation of autonomy to the local level of the organization. Most significant is the absence of shop floor group work supported by the system. The implementers of *BaaN* represent mostly medium to large manufacturing companies (around 500 employees) with one geographical location. There is not much further centralization possible in their organization. Within this community there are some examples of delegation. Finally the *Navision* community is the largest by number of companies, but also the smallest in size. The manufacturing companies within this group seem to celebrate a much more substantial praxis of autonomy supported by the system. Several companies in the sample thus use the systems to support shop floor teams.

The community constrains and enables the intra-organizational political process of implementing the ERP system. It is more important than ever to create policy processes with broad participation in order to enable development of a governing strategy of how to use the systems. Otherwise technology management will suffer from the 'power of default', the unwillingly forced use of myriad pre-set parameters, instead of exploiting what should have been the room for manoeuvre.

Management of technology, the IS-specialism and organization studies continue to prefer to take the organization and the single software system as the analytical unit when they discuss implementation of technology. The point of departure for this contribution is that this analytic unit becomes increasingly misleading when the technology is ERP software packages. These systems are a result of a mass production of software, which changes radically the options for the enterprise actors. Partnership with the software houses 'behind' such software is broadly speaking out of the question. This new situation challenges management. Priorities have to be made, the enterprise actors have to formulate basic policies and the limited resources for reshaping the systems should be set in.

Drawing on a combination of organizational politics and sociology of technology the view taken on ERP is that it is a political programme for change. An inspiration in the understanding of the complicated role of technology in political processes is arguments mobilized by Orlikowski for an intertwined processual and structural view of social processes and technology (Orlikowski 1992). The context used here is the development, implementation and use of ERP software systems within the scope of manufacturing in Denmark. These systems rely on a bundle of other technologies, including systems for telecommunication and basic functions (hardware of various types), but the particular focus taken here is on the software side of the systems. The systems discussed are *SAP R/3*, *BaaN IV* and *Navision XAL* (the latter was developed originally by a Danish software house, which was bought by Microsoft in spring 2002).

The chapter opens with an outline of the main theoretical elements. First, the political process perspective on organizational change and the concept of mass-producing communities is described. Second, case study material from Danish manufacturing is used as a vehicle for discussing the role of communities in closing the flexibility of ERP. This section includes a characterization of the ERP technology as a political programme. The case material proceeds at two levels: the community level and two enterprise case studies. Finally, the conclusion discusses the implications of the role of communities.

The political process approach to organizational change

Based on empirical grounds, it is my contention that the everyday life of management of information technology is full of ambiguities, variants of technology, different actors seeking to turn decisions in other directions and other social phenomena. This points at a conceptualization of management of information technology as a political process. This is done elsewhere (Koch 1998), drawing on organizational politics studies, organizational sociology (e.g. Pfeffer 1981; Pettigrew 1985; Midler 1993; Knights and Murray 1994) and the sociology of technology (e.g. Callon 1987; Law 1991; Latour 1997; Law and Hassard 1999). These positions share a number of understandings of the processes of negotiation and coalition building, but differ in their perception of the role of technology. The conceptualization is briefly sketched below as a frame for the focus here on the role of technology as a political programme in political processes.

Political processes may be understood as a combination of a political programme and a coalition-building process. These two elements dynamically intertwine. Thus, when enrolling actors in a coalition, it is likely that the political programme changes (as demonstrated in numerous of Latour's writings, e.g. Latour 1987). Political programmes emerge from the intentions of the actors in the setting. It may be described as a merger of intentions, joining and directing the coalition in a specific direction. It is likely that the coalition and the political programme continue to be unstable and under negotiation, since it frequently unites actors with rather different intentions and interpretative worlds. On the other hand, this social glue is of central importance for the programme in order for it to be workable. Participation in a coalition promoting a programme is likely to change the actors themselves (actor network theory (ANT) speaks of translation of interests). The merge of intentions frequently leads to the shaping of relatively few obligatory points of passage, simplified elements of the programme that act as representatives of the programme's larger agenda. The arena for the political process is likely to be different from the isolated organization. It will frequently include external actors such as IT suppliers and management consultants.

The structured inequalities between actors, the societal embeddedness and the limited scope for voluntary decision-making and agency within a specific organizational context are important features of the concept of political processes (see also Knights and Murray 1994). In other words, interest in the process leads to a renewed interest in structural constraint and enablement. Structures are viewed as omnipresent and are thus a direct part of the process.

Studies in the sociology of technology (STS) can repair a major shortcoming of the 'pure' political process approach as it appears in organizational politics literature: the relative absence of technology as analytical category. Studies informed by SCOT and ANT (Bijker 1995; Latour 1987) offer a concept of technology in which the technical and the social are viewed as a socio-technical ensemble, a seamless web. Technological development is seen as co-shaping the technical and the social (see Chapters 1 and 2, this volume).

Political programmes

A political programme, in its seminal forms, is a means of thinking about the content of change and how to obtain it. The emergence of a political programme usually results from some kind of re-adaptation of old thinking shaped in a new way, as also pointed out by several observers (e.g. Earl and Khan 1994). It reflects some of the basic features of the context in which it operates. This embeddedness is a necessary precondition and enabler for a new programme. On the other hand, a change programme is usually different in its content from the contemporary context. This difference is part of the mechanics of the process (see Buchanan and Storey (1997); Buchanan and Boddy (1992) for a discussion of the relationship between change programmes and the context of the organization).

Active intentions of actors are one part of the process leading to a political programme. According to the processual approach, organizational life is characterized by a multitude of different actors with different intentions in the form of understandings of problems and solutions, different rationalities, knowledge, interests and experiences. All are brought into a negotiation process, which may lead gradually to common intentions. These intentions are transformed into a political programme. In this chapter the focus is on the elaborate and explicit end of a range of programmes: the commodified management concepts such as business process re-engineering (BPR) (see Chapter 10, this volume). Commodified technologies such as the intranet (see Chapter 3, this volume), FMS, CAD and ERP are examples which contain a technology or an IT system as core. Both ERP and BPR exist in a number of variants and are variously articulated by consultants and others (Koch *et al.* 1997, 2000).

Critical readings of large-scale political programmes such as management concepts are numerous. Midler (1986) and Huczynski (1993) argue that such a political programme usually contains a piece of theory, some experiences made abstract and general. A concept will typically contain a diagnosis of problems and some suggestions for solutions. Furthermore it will contain methods for analysis and suggestions for change processes, change management. More implicitly, the concept contains a view on man and the organization, and, finally, some practical experiences, the presence of which are often a precondition for any presentation.

The emergent characteristics of the programme will be discussed in the cases outlined below. It turns out that the programme is not stable, but has changed as part of the linking up of actors. Through negotiation, the programme has changed and developed. The initial take-up of, say, BPR or ERP in an organization thus normally leads to a changing agenda over time that in the end leads to other changes in the organization than the ones foreseen.

Mass-producing communities

The concept 'mass-producing communities' is used to grasp a constellation of a technology and an adjacent actor grouping. The concept was originally empirically induced to describe the new situation that Danish medium-sized manufacturing enterprises entered during the 1990s. Certain software houses producing ERP systems managed to develop a mass-production situation, where software packages are sold through a large network of sales offices, so-called 'value-added resellers' (VARs), implementation consultants and so on. Examples include SAP, *BaaN*, Sage, MFG: Pro, Peoplesoft and *Navision*. These actors have a common central knowledge and meaning-assigning object, the ERP system. The systems and the routines attached to implementation, customization and development of the systems are a glue of this particular social constellation. The members of the communities are individuals and organizations such as customer enterprises, suppliers, management consultancies, education institutions and others. Labour is exchanged throughout the community and the software houses often have courses, training and certification programmes for professionals working in the community. In this sense it shares features with professional associations (Swan 1997). The courses, moreover, create links across the community (Hansen 1998). The groupings are, however, relatively loose and members of the community are in perpetual competition (consultancy companies and VARs). It would thus be misleading to talk of a network since there are social barriers within the community and competition.

The concept differs from communities of practice and from socio-technical ensembles (Bijker 1995; Wenger 1998) although there is considerable overlap. Communities of practice (c-o-p) differ in two major ways. First, the central role of a technology (here the ERP system in the mass-producing community) is downplayed in Wenger's conceptualization of c-o-p at least. He gives preference to more individual meanings and 'smaller' artefacts (Wenger 1998: 83). Second, c-o-p leans too much towards a harmonic understanding of a joint enterprise for the community; Wenger thus notes of 'joint enterprise' as a feature of c-o-p that it is 'not just a stated goal, but creates among participants relations of mutual accountability to become an integral party of the practice' (Wenger 1998: 78). In contrast to this, the mass-producing communities are characterized by inherent, even structural internal conflict and tensions.

The technological ensemble gives a larger role to technology in its conceptualization of what holds the social group together. In Bijker's thinking technology and the social are symmetrical. This comes closer to the mass-producing communities than do the communities of practice. Bijker however assigns too much stability to the socio-technical ensemble. The mass-producing communities in contrast are in high-speed development and in hyper-competition with each other. They evolve and dissolve over relatively few years.

On the other hand, all three concepts share the understanding of an exclusion mechanism as the other Janus head of the glue features. Communities can be fortresses, but they can also be open to external actors (Wenger 1998). Moreover, the three concepts share the understanding of the importance of daily work practices and social interaction as gluing the socio-technical unit together.

ERP: Information Technology as a political programme

The focus of this section is the empirical material. The section develops as follows. First, some results are given from a critical reading of ERP systems in the form of a discussion of the content of the systems. Second, a presentation of the SAP community is offered as an example of the mass-producing community. Third, a presentation of the analysis of three communities in Denmark, and their impact on how ERP was implemented, is given. Fourth, two in-depth case studies of a SAP and *BaaN* implementation are discussed.

Method

This section draws mainly on two types of empirical material. First, twenty-five cases from Danish manufacturing were studied between 1996 and 1999. Most were covered by one-day visits, whereas four were followed over a period and visited several times. The ERP systems studied were *SAP R/3*, *BaaN* and *Navision XAL* (originally a Danish software package). Second, two *ex-ante* process follow-up studies were carried out on the developing relations between two Danish manufacturing enterprises, Olsen Produktion (Olsen), Jensen Manufacturing and two ERP software suppliers, SAP and *BaaN*, other suppliers and a management consultant (Olsen and Jensen are pseudonyms). Olsen and Jensen are medium-sized manufacturers with a series of typical functional departments. The studies were carried out between 1995 and 2000, and are a result of two years of intensive fieldwork, supplemented with more extensive co-operation covering five years. A palette of methods was used which justifies labelling the empirical work ethnographically informed phenomenological fieldwork. It comprised participant observation, meetings with enterprise representatives, regular telephone conferences with the project manager, diary management and semi-structured interviews.

The mass producing communities are a widespread and complex phenomenon, being to some extent a global phenomenon. The communities have been followed predominantly through the perspective of Danish manufacturing companies and events, press and books from professional associations and so on in a Danish context. I have participated in numerous seminars arranged by ERP vendors, professional associations and others in Denmark, Scandinavia and Germany in the period from 1993 to 2000. This means that some information on players is specific in time and space, since it refers to this period and to the arena in Denmark.

The content of enterprise resource planning systems

The IT systems in focus here are 'Enterprise Resource Planning' systems (ERP-supersite 2000). The development of planning and control software for manufacturing enterprises is accompanied by continual change in the notions used to name them. A brief outline of the main development is offered below, along with some of the names.

Enterprise Resource Planning systems (ERP) is a merger of a number of visions of control and accompanying routines and practices. However, one can point to mainly three visions (or *leitbilder*: see also Hamacher, in Heilige 1996) of controlling a manufacturing enterprise:

1 Economic vision: The enterprise as a financial entity with economic flows.
2 Logistics vision: The enterprise as material flow.
3 Information vision: The enterprise as information system and flow.

A number of systems (one Danish reference says 250: Jacobsen 1996) are present on the Danish and the related international scene. The SAP system *R/3* is a proper template. The main modules of the system in version 4.6, 2000 are as follows:

FI: Financial management
CO: Controlling
TR: Treasury
MM: Material management
PP: Production planning
PM: Plant maintenance
SD: Sales and distribution
HM: Human resources
QM: Quality management
PS: Project-management module
WF: Workflow
EC: Executive information system
BIW: Business information warehouse

As may be noted, a series of departmental and functional elements, in addition to the three main visions mentioned above, have been included in the system. This applies to sales and maintenance, for example. The three main visions are incorporated in the following way: the first three modules are related to the economic vision of control. Within the next two modules of MM and PP are the main parts of the MRP II realization. The information vision is not visible in this list of application modules, but is realized through an underlying database, development tools and so on. Each application module has a number of sub-modules. One example of this is the product configuration facility discussed below. In *R/3*, this facility is part of the production-planning module (PP). Through the implementation process the clash between routines of the enterprise and the suggested processes within *R/3* (around a thousand) leads to a number of 'quiet battles' between the two or several logics. How this turns out is heavily dependent on intra-organizational politics, and it produces a number of variants which develop as a result of different micro-political choices (see Koch 1999):

- Choice of modules (including industry specific).
- Choice of parameters.
- Choice of user profiles.
- Choice of additional programming.
- Choice of reports.

The overall design of *R/3* is a model of the overall enterprise and corporation. This model could be interpreted as an integration and centralization push. The integration push transcends single enterprises and encompasses whole corporations with rows of divisions and units. Davenport notes that ERP is indeed a push in the direction of full integration of previously separate functions and entities, even in cases where separation may be competitive (Davenport 1998; see also Williams and Edge 1996). The integration within *R/3* however is not so tight that enterprises cannot get a running system out of a more restricted version with only partial integration. In fact, the majority of business processes in the system are designed to operate within a main module.

Other levels of design include business processes and user profiles. Around a thousand recommended business processes are offered in the system. The policies of adopting the standard SAPs processes lead to significant savings in resources in the implementation process and to what may be termed '*the power of default*': Although in principle it is possible to design user profiles, parameters and business processes in another way than SAP suggests, the enterprise actors are likely to take the short cut of using suggested parameter settings and choices.

The SAP community in Denmark

The Danish department of SAP has around 150 employees and has been growing rapidly since 1993. The Danish customers cover all sectors and include the 'heavyweight' players in Danish manufacturing: Danfoss, Novo, Carlsberg, B&O, Q8/DK, Danisco, ØK, APV, Lego, Statoil and Grundfos. Nevertheless, it is important to emphasize that the Danish community is a 'remote' part of the global SAP community, which encompasses a major development unit in Germany of around 20,000 employees and about the same amount of sold systems.

Companies such as Danfoss and Lego were early adopters of *R/3* in Denmark. Such first movers play an important role in building the mass-production community. SAP has been spreading from such major manufacturing players downwards into medium-sized groups and into other sectors such as the public and finance. In 2000 there were more than 150 customers (see Table 4.1).

The *SAP R/3* community is underpinned by a SAP–Denmark education and training programme which, apart from customer training, also encompasses programs for consultants, programmers and updates as well as basic programs. Moreover, SAP offers, through the so-called 'Nordic Academy' to

Table 4.1 The SAP *R/3* community in Denmark

SAP
SAP–Danmark
Management Consultancy (*see below*)
Educational institutions
Approximately 150 customers
Management consultancy
Andersen Consulting
Aston IT
CAP Gemini
CSC Danmark
Deloitte & Touche Consulting
ECsoft Danmark
Ernst & Young
IBM
KPMG
Origin/Danmark
PA Consulting
PriceWaterhouseCoopers
Corebit
Siemens Nixdorf
EDS
EDB-Gruppen
Datacon
PCA
Kommunedata

develop consultants whom the companies can 'take over' afterwards (Hansen 1998). This community has a developed set of understandings of how the *R/3* software should be configured and customized. This becomes apparent when enterprise implementations are studied. *R/3*, the software package, is a central part of the community and serves as a glue to keep the grouping together. The major part of the development of *R/3* is situated outside the Danish community. Table 4.2 outlines the different layers of the shaping and development of *R/3* since 1990.

SAP usually sets up partnerships when it develops the basic software packages. The Danish part of the chemical group Kemira participated in the development of the PP-PI module in 1993 to 1995 as partner in the group developing this particular module. Similarly the Danish Jydske Bank has participated since 1998 in developing the suite of modules for the finance sector. The majority of Danish companies and groups, however, are forced to accept the basic architecture and basic functionality of *R/3*. SAP strategy in Denmark at the beginning of the *R/3* period was, moreover, to find enterprises that fitted 90 per cent to the system (Koch 1994).

Table 4.2 The shaping and development of *R/3* since 1990

Development of SAP R/3, *space*	*Shaping options, occasions*	*Shaping options, continual*
'Global' SAPs department of Development Walldorf	1993 development of *R/3* basis system, all modules 1994–1999 new versions of the system (1997–1998 finance sector system) 1998–* public sector system New basis system *R/4**	1999 development of version 4.6 series Development of version 5.0*
National SAP–Danmark	1994 Danish version of *R/3* 1998 Dansk version of finance system 1998 Danish version of human resource module	1999 smaller changes in specific Danish parts
User association FSD (For. Af SAP-brugere i Danmark)		Input to SAP–Danmark
Consultants	At first enterprise contact: implementation model, SAP Competence, organization model, management concept(s)	During implementation: Consultants in enterprise
Groups/enterprises	Implementation: Choice of main modules Choice of sub-modules Choice of concepts of control Configuration: choice of parameters User profiles	During use Further modules and sub modules Use of ABAP and report generator Customization
The workplace	Possible participation in implementation	Super-user support Small customizations Appropriation of user screen Reports and lists

Note
*Indicates that the event is anticipated but has not yet occurred.

The general results from the sample cases: mass-production communities

The case studies of the three communities around the ERP systems *SAP R/3*, *BaaN* and *Navision XAL* show that even though each enterprise case is unique in principle, one can point to a number of similarities between them. This implies that a key shaping arena for the political programme of ERP is

Table 4.3 Change programmes in twenty-seven ERP implementations

Change/system programme	SAP	BaaN	Navision XAL
Substitution of IT	2	4	8
BPR before ERP	4	2	–
Redesign after ERP	2	1	–
Other	1	2	1

actually these communities, consisting of software houses, vendors, consultants, education and training units, customers and others.

Table 4.3 shows the main programmes in the enterprises. The characterization is rough and ready, several enterprises fitting into more than one category. Substitution of IT occurs when the old system is replaced by the new with no prospects for change (BPR before ERP does not exclude the fact that you redesign afterwards, for example).

The *R/3* implementations are characterized by the most ambitious programmes of change in *the overall design of the business*. Business process re-engineering, centralization and (modest) virtual integration of a number of geographically spread units are central features. At the same time the *R/3* community is characterized by a relative lack of decentralized organizational change. The *BaaN* and *Navision XAL* implementations on the other hand are mixtures of different organizational changes: decentralized, status quo or more centralized versions. The *BaaN* enterprises are medium-sized 'one site' enterprises with fewer possibilities of virtualization. Finally the *Navision* enterprises are small enterprises with typically fewer than 500 employees, and the implementation is characterized by a status quo replacement of former systems, and some decentralized models (supporting group work in production, for example).

Across the three communities the enterprise choices within the *design of major business activities* are a combination of common choices and unique single enterprise choices. The implementations are grouped into three variants in each of the three system communities characterized by different choices of modules. The common choices encompass the economic system, which is the basis for the vast majority of cases, whereas the unique enterprise choices at major business activity level, for example, are the choice of the process module in *BaaN*, which is done by one case enterprise only (Table 4.4).

Table 4.4 Major variants of the three ERP systems

System	SAP	BaaN	Navision XAL
Major variants of choice of modules	• Economy only • Full package • Industry solution	• Logistics only • Full package • Integration of group	• Economy only • Economy and logistics • Full package

The *design of user profiles, setting of parameters in business processes* and other micro-political aspects represents a vast variety across the cases. As will be discussed below, the micro-design becomes detached from the major political programme visions about the change, or represents areas where the programme does not express needed change. For example, there are both narrowly and broadly designed user profiles in all three communities. When it comes to the support of shop-floor activities, there are marked differences between the SAP community and elements of the two others. One finds within the *BaaN* and *Navision* communities a large number of examples of shop-floor workers using the systems to release production orders, do fine scheduling and report on production figures. This user profile is rare in the SAP context.

The mass-producing communities play a role in the negotiation of some elements of flexibility in the software systems, such as the lack of decentralized models in the SAP context. On the other hand, the internal variety is considerable and there is clear room for intra-organizational politics within the community, as the following two longitudinal cases also demonstrate.

The SAP case: BPR and ERP as prescribed medicine

Jensen is a process manufacturer. It was formed through a merger of three companies in the 1980s, each having a network of sales and distribution units attached to the main factory. Before the change project, the group consisted of a distributed network of units. The management group was equally diverse. The shop-floor worker shop stewards in the three factories dominated union activities in the group, while the shop stewards in distribution and administration were inactive, consenting silently to developments. Relations with management were adversarial to a certain extent, co-existing with co-operation.

The change process fell into three main phases:

1 The BPR analysis.
2 The BPR implementation and the configuring of the *SAP R/3* system.
3 The implementation and use of *SAP R/3*.

A management coalition from the group decided on a BPR and *SAP R/3* project in 1995. The BPR analysis carried out in the following year involved the formation of a joint coalition consisting of the external consultants and a major project group of sixteen members. An initially broad and relatively general political programme with the headlines of BPR, 'clean sheet', a focus on cross-cutting processes and the like evolved into a more substantial programme during the analysis. The analysis pointed up three 'core value creating' business processes and seven support processes, along with around eighty suggestions for the reorganization of the group. Through the BPR analysis it emerged that the group wanted to achieve a virtualization of the sales and ordering process, wage administration, purchasing, and to some extent production. Manufacturing resources were to be co-ordinated across the different locations.

During the BPR analysis, top level and project management were slightly detached, using this distance to stall shop steward enquiries about the development of the project. While one group in the project organization was analysing the production facilities in detail, aiming at stronger co-ordination and rationalization across the three main factories, shop stewards were told that production was not part of the project work. This kind of machiavellian strategy occurred on several occasions throughout the project. As part of the BPR implementation process sales employees were dismissed, but could apply for a limited number of newly designed jobs. The distribution network was declared outside of the project by top level management one month before a major outsourcing of distribution tasks and two months before a major reorganization of the distribution network, a process leading to major changes in work and a few dismissals.

Shop stewards reacted to this management strategy with unease in the BPR analysis phase. While they tried to maintain a co-operative attitude, they were receptive to external input by their union and the researchers, arguing that they needed their own expertise and resource development. This was achieved one year later, when the analysis was almost complete.

In the BPR implementation phase substantial reorganization took place in the administrative and distribution units of the group. Finance, sales, ordering processes, wage administration and purchasing were substantially reorganized. The administrative changes consisted of the concentration into one unit of a number of processes and the linking of them with the other units through IT. Even during this phase management continued to maintain that production facilities would not be changed. However, this position changed gradually as the configuration of *SAP R/3* became more and more detailed, a process which normally needs support from employee representatives; that is, providing detailed knowledge on the business processes to be configured in the system. This was carried out in an informal fashion, with direct contacts between the SAP project group and individual employees.

When *SAP R/3* was implemented, a number of changes occurred in production work. Cross-cutting processes were realized internally in two of the three plants, leaving the third behind in the first round. The cross-cutting processes included linking goods-receiving facilities with purchasing, the distribution of schedules and changes thereto, linking the stocks of the entire group into one stock administration system and so on. The informal and detached configuration and training strategy led to different levels of use and access for shop-floor workers.

A number of administrative processes were centralized into one unit, with linkages to all the other units. Wage, time and attendance administration relied on local employees registering in co-operation with human resource staff. Purchasing was centralized, leaving local employees and first line managers with restricted autonomy in some aspects of purchasing. Second, different units were integrated electronically inside and outside the factories – for example, sales was linked to order handling, scheduling and machine/shop units. Finally, the planning of production now involved viewing all the production facilities as a resource for production, thus combining geographically spread units to a greater extent than previously. The production process was executed by balancing capacity across the factories and transporting semi-finished products between sites. The stock of finished goods at different distribution units was viewed as a whole, and thus stock was transferred between units according to sudden changes in requirements. Technical problems occurred, especially in the realization of the sales order process across geographical units. This meant months of struggling with the system. The group claimed that it lost £2m in turnover during these months.

This case demonstrates the realization of a relatively successful 'high order' political programme through the merger of three factories into one manufacturing network and one supply chain. The overall design of the enterprise was changed according to the programme developed in the analysis phase, mainly through organizational means, but also by the underpinning configuration of *R/3* and the choice of main and sub-modules. Within this frame occurred local political behaviour in the shaping of the functions within the departments of the organization. These local processes shaped user profile access to certain sessions and soon seem to have been influenced by local considerations for local human resources, work organization and the like, mainly with the aim of maintaining the status quo.

The BaaN *case: the multi-front operation*

Olsen is characterized by discrete high-volume production combined with a large net of suppliers. Industrial relations are characterized by high trust in the main factory. The sales function occupies a dominating position, and the current CEO stems from sales. The change coalition, however, is predominantly embedded in the manufacturing departments. The participating consulting companies are given the pseudonyms Klein and Johansen. Two ERP vendors competing with *BaaN* are labelled SYSCOM and DYNAMPS.

The change process fell into three phases:

1 Preparation and analysis, lasting twenty-three months (including a pause).
2 Configuration, lasting fifteen months.
3 Implementation, lasting eighteen months, including a twelve-month low activity period due to performance problems, change in consultants and a pause induced by a global *BaaN* crisis.

The change programme was an amalgam of the intentions of the participant actors joining and directing the coalition in specific directions. At Olsen, the initial vision was to substitute the old ERP system with a new one, but during the first year of planning it turned into a strategic project which formulated visions on organizational change as well. There is an element of emergent process in the way the strategy project came about. The enterprise was sold to another group and the *BaaN* project was halted for six months. It was during this pause that Olsen hired the consultants Johansen to conduct an analysis of the possibilities for cost-cutting and changes in business processes. This project involved a number of managers and employees at Olsen, and led to a common formalized strategy which had clear cross-cutting elements. The strategy maintained a split between sales and manufacturing and the implementation of *BaaN* was a central element. Thus the change programme after one year of planning consisted of an intention of organizational change in the form of a set of loosely co-ordinated change/cross-cutting processes and the technological/organization change in the ERP programme.

During the initial choice of ERP there was a focus on order handling for several reasons. ERP systems are rather complicated, and certain elements were especially important in order to glue manufacturing and sales together in the change coalition. During the process the participants necessarily carried out simplification ('shaping of obligatory passage points': Latour 1987). The participants thus relied on 'popular' interpretations of the main areas of the systems (accounting, logistics and IT). One system, for example, was interpreted as strong in accounting but weak in its IT features. In combination with these simplifications, the coalition had a close look at a few facilities of the systems, especially the product configurator facility. The following differences were found between the three main areas of interest:

- *BaaN IV* has a built-in product configurator.
- DYNAMPS should be coupled with third-party software.
- The supplier of SYSCOM offered joint development of a product configurator.

The product configurator acted as a key obligatory point of passage for the coalition, as this feature represented the realization of its vision for the scope of the system: the aim was to integrate upstream along the sales order process all the way out to the first encounter with clients. The systems were, however, interpreted quite differently by the enterprise actors, so sales did not think that *BaaN* would offer an attractive substitution for the existing IT configuration. The MRP II part of the systems was examined in detail, and this led to an internal coalition consensus that pointed at *BaaN* and Klein, the consultants that should implement.

It took several years to establish a wider scope for the IT solution and a broader supporting coalition. The original vision – full integration along the order generation process – was only part of the company's future plans even by the summer of 2000.

Throughout 1997 to spring 1999 the configuration and customization process absorbed most of the energy of the *BaaN* project group. The way in which the project was organized and the amount of work invested in technical issues meant that the organizational strategy elements almost disappeared from the agenda. In the configuration of certain business processes, however, cross-cutting elements were realized. During configuration the power of default was exercised in a surprising way: the finance department decided to configure *BaaN IV* exactly as in the previous system.

In a sense as a secondary but related side theme, the competition between a technological and an organizational change programme started in 1999 after almost a year of parallel existence. During 1999 to 2000 it became evident that management needed to reconcile the vision of full control of the assembly process, which was embedded in the ERP coalition, with the idea of a 'teams in production' coalition. Teams were introduced in spring 1999, and it emerged as an element of team-working that shop-floor workers could and should be users of ERP. Although this remained a latent conflict, it challenged the high-trust stability in Olsen. Compromises were struck while the system was halted due to performance problems in autumn 1999. So, while social distance was created temporarily to realize a control system for assembly, it was later dissolved again when the configuration was over and the system stabilized. In addition, the halt of system implementation indirectly created resources for this compromise process to occur.

Conclusion

The main enterprise sample largely confirms the concern that the 'flexibility in principle' of ERP systems is 'gone' even before the enterprise starts an implementation process – a feature that points to the role of the communities of practice. The mass-producing communities (i.e. the network of ERP vendors, consultants and other customers) built up practices that downplay elements of flexibility in the software systems, such as the lack of decentralized models in the SAP context. On the other hand, internal variety on all the three levels discussed (i.e. the overall design, the major business activities and the micro-level) is considerable.

The *R/3* implementations are characterized by the most ambitious programmes of change in *the overall design of the business*. BPR, centralization and some virtual integration of a number of geographically-spread units are central features. The *BaaN* and *Navision XAL* implementations on the other hand are mixtures of different organizational changes: decentralized, status quo or more centralized versions. The *BaaN* enterprises are medium-sized, one-site enterprises with fewer possibilities for virtualization. Finally, the

Navision enterprises were characterized by a status quo replacement of former systems and some decentralization models.

Across the three communities, the enterprise choices within the design of major business activities are a combination of common choices and unique single enterprise choices. The common choices encompass the economic system, whereas the unique enterprise choices at, for example, major business activity level, are the choice of the process module in *BaaN*.

The design of user profiles, setting of parameters in business processes and other micro-political aspects represent a vast variety across the cases. The micro-designs become detached from the major political programme visions about the change, as well as the pre-configured processes in the ERP systems. There are also both narrowly and broadly designed user profiles in all three communities, which represent consent to and break from the ones proposed by the ERP vendor. When it comes to the support of shop-floor activities, there are marked differences between the SAP community and elements of the other two.

Within two of the communities, SAP and *BaaN*, the longitudinal cases have discussed enterprise politics. The SAP case demonstrated how a 'prescribed medicine' in the form of a 'headline' management political programme of BPR and *R/3* developed into a more explicit change programme, which was then realized relatively successfully. External consultants played a central role in this process, especially the 'high order' political programme of organizational change, merging the three factories into one manufacturing network and one supply chain. The overall design of the enterprise was changed according to the programme developed in the analysis phase, mainly through organizational means, but was also underpinned by the choice of main and sub-modules in *R/3*. The resources mobilized in the group were concentrated in the large project group and by hiring external consultants. On the other hand, production managers, shop-floor workers, shop stewards and some administrative staff were largely excluded. Social distance was thus used to create a disclosed arena for the analysis and implementation of changes. The technology, *R/3*, was used to underpin these major changes by the configuration of modules and sessions, especially in the sales, purchasing and finance modules. Within this frame, however, is to be found local political actions in the shaping of the functions within the departments of the organization. These local processes shape user profile access to certain sessions and so on, and seem to have been governed by local considerations for appropriate functionality, local human resources, work organization and the like, and mainly with the aim of maintaining the status quo. The processes were, moreover, halted by a lack of resources (mainly time) at the end of the configuration process. In this sense the power of default played a certain role. The local processes did not lead to a break with the SAP community embedded praxis of not developing decentralized models, and neither did the general design realized in the group.

The *BaaN* case demonstrates a multi-faceted exercise for the management coalition. While trying to enrol sales for the full ERP programme, assembly employees and shop stewards were held at a temporary distance, while realizing a control and scheduling system. On the one hand, the management coalition, with its strength in production, tried to enrol the sales function in an enlarged coalition with a 'full ERP' as a political programme. The programme encompassed a redesign of the overall business, and enabled further integration through cross-cutting processes. Despite several attempts, using logistics consultants, *BaaN*, the product configurator and IT consultants, and the broader alliance, the programme was not fully realized. The ERP agenda seems rather too strong for this purpose, and is underpinned by external communities. Nevertheless, the programme as a *'leitbild'* not only produced consent but also resistance, and the 'hardness' will need to be reshaped to enable consent by certain actors, thus giving the process an emergent character. There is thus continual debate on the overall business structure, especially the relationship between sales and production. On the other hand, the competition between the technological and organizational change programmes began at the end of the 1990s, after almost a year of parallel existence. Although it remained a latent conflict, it challenged the high trust stability of Olsen. New compromises were made, however, while the system was halted due to performance problems. So, while social distance was created temporarily to achieve the control system for assembly, it was later dissolved again when the configuration was complete and the system stabilized. What is more, the halt in system implementation indirectly created resources for this compromise process to occur.

Both cases thus demonstrate how technology, *R/3* and *BaaN IV*, and its supporting communities, can overshadow intra-organizational management politics. ERP, BPR and total logistics in the three cases did not represent a break with the external communities of practice (it should be added that in the Olsen case some of the managers were also active in professional associations). The former three-factory structure of Jensen was thus changed into one much more unified organization. Integration in cross-cutting processes and enhanced control over assembly was almost realized in Olsen. However, in both cases the 'external community policy' was controlled jointly by the community and the internal element of the management coalition. One should not picture the enterprise actors as victims. Rather, the process reconstituted the hegemony and adversarial relations between the management coalition and the shop-floor in the Jensen case, and challenged the high-trust stability in the Olsen case.

Moreover, it should be emphasized that while some features were frozen before, others thawed out, and could be configured in micro-political processes. In the analysis it has been demonstrated that the overall business structure, business processes and micro-level elements can all be reshaped, thus leaving the question of the degree of 'hardness' open to empirical investigation. It has been suggested that the communities around ERP systems should be incorporated into the analysis. In this way, in conceptualizing the freezing process as a long-term, multi-actor process with occasions and spaces for reopening issues (Clausen and Koch 1999), the 'thoroughgoing interpretative' position (see Chapter 2, this volume) can be left behind. It is more important than ever to create policy processes with broad participation opportunities embedded in (for example) professional associations and/or public policy, in order to enable the development of a 'governing strategy' for how to use the systems. Otherwise technology management will suffer from the 'power of default', the constraint of the unknowing/unwilling use of myriad pre-set parameters, rather than the exploitation of choices for manoeuvre.

References

Alvesson, M. and Willmott, H. (1996) *Making Sense of Management. A Critical Introduction*, London: Sage.

Badham, R. (1995) 'Managing sociotechnical change: a configuration approach to technology implementation'. in J. Benders, J. de Haan and D. Bennett (eds) *The Symbiosis of Work and Technology*, London: Taylor & Francis.

Bancroft, N.C. (1996) *Implementing SAP R/3. How to Introduce a Large System into a Large Organization*, Greenwich: Manning.

Bijker, W. (1995) *On Bakelite, Bulbs and Bicycles*, Boston, MA: MIT Press.

Buchanan, D. and Boddy, D. (1992) *The Expertise of the Change Agent,* New York: Prentice Hall.

Buchanan, D. and Storey, J. (1997) 'The creatively orchestrated performance', Working paper, Loughborough Business School.

Callon, M. (1987) 'Society in the making: the study of technology as a tool for sociological analysis', in W. Bijker *et al.* (eds) *The Social Construction of Technological Systems*, Cambridge, MA: MIT Press.

Callon, M. (1991) 'Techno economic networks and irreversibility', in J. Law (ed.) *A Sociology of Monsters: Essays on Power, Technology and Domination*, London: Routledge.

Clausen, C. and Koch, C. (1999) 'The role of occasions and spaces in the transformation of information technologies'. *Technology Analysis and Strategic Management* 11, 3: 463–482.

Clegg, S. (1989) *Frameworks of Power*, London: Sage.

Clegg, S., Hardy, C. and Nord, W. (1996) *Handbook of Organisational Theory*, London: Sage.

Davenport, T. (2000) *Mission Critical*, Harvard University Press: Cambridge

Earl, M. and Khan, B. (1994) 'How new is Business Process Redesign?'. *European Management Journal* 12, 1: 20–30.

ERP supersite (2000) http://www.erpsupersite.com.

Fleck, J. (1993) 'Configurations: crystallizing contingency'. *International Journal of Human Factors in Manufacturing* 3, 1: 15–36.

Fujimora, J. (1992) 'Crafting science: standardized packages, boundary objects, and translations', in A. Pickering (ed.) *Science as Practice and Culture*, Chicago, IL: University of Chicago Press.

Grint, K. (1995) *Management – A Sociological Introduction*, Cambridge: Polity Press.

Grint, K. and Woolgar, S. (1997) *The Machine at Work: Technology Work and Organization*, Cambridge: Polity Press.

Hansen, H.A.B. (1998) Personal communication by the author with H.A.B. Hansen, SAP course participant.

Hanseth, O. and Braae, K. (1998) 'Technology as traitor: emergent SAP infrastructure in a global organisation', in R. Hirscheim, M. Newman and J.I. DeGross (eds) *Proceedings of 19th International Conference on Information Systems (ICIS)*, Helsinki: ICIS.

Harvey, D. (1996) *Justice, Nature & The Geography of Difference*, Cambridge, MA: Blackwell.

Haugbølle Hansen, K. (1998) *Den sociale konstruktion af miljørigtig projektering*, Ph.D. thesis, Institut for Planlægning. Lyngby: DTU.

Heilige, H.D. (1996) *Leitbilder auf dem prüfstand*, Berlin: Edition Sigma.

Henderson, K. (1998) 'The role of material objects in the design process: a comparison of two design cultures and how they contend with automation'. *Science, Technology & Human Values* 23, 2: 139–175.

Huczynski, A. (1993) *Management Gurus – What Makes Them and How to Become One*, London: Routledge.

Jacobsen, L.C. (1996) 'Forstå MRP II'. *Effektivitet*, 4/96, Copenhagen.

Knights, D. and Murray, F. (1994) *Managers Divided*, London: John Wiley.

Koch, C. (1994) *Teknikere og Produktionsstyring*, Teknisk Landsforbund: København.

Koch, C. (1997) 'Production management systems – bricks or clay in the hands of the social actors', in C. Clausen and R. Williams (eds) *The Social Shaping of Computer-aided Production Management and Computer-integrated Manufacture COST A4*, EU Commission.

Koch, C. (1998) 'Management of technology. Coalitions, networks, political processes', Working paper. Mimeo. Bremen University, ARTEC.

Koch, C. (1999) 'SAP R/3 – an IT plague or the answer to the Tailors dream?' *Proceedings*, 2, *PICMET*, Portland International Conference on Management of Engineering and Technology, Oregon.

Koch, C. (2000) 'The ventriloquist's dummy? The role of technology in political processes'. *Technology Analysis and Strategic Management* 12, 1: 119–138.

Koch, C., Manske, F. and Vogelius, P. (1997) 'The battle for the soul of Management Denmark – the shaping of the Danish versions of Business Process Re-engineering', in E.H. Bax (ed.) *Proceedings from 'Management at a Crossroads'*, Groningen University, Faculty of Management and Organization, pp. 360–380.

Kotter, J.P. (1996) *Leading Change*, Boston, MA: Harvard Business School Press.

Latour, B. (1987) *Science in Action*, Milton Keynes: Open University Press.

Latour, B. (1997) 'On Actor-Network Theory – a few clarifications', *Soziale Welt.*

Law, J. (ed.) (1986) *Power, Action and Belief*, London: Routledge.

Law, J. (ed.) (1991) *A Sociology of Monsters: Essays on Power Technology and Domination*, London: Routledge.

Law, J. and Callon, M. (1992) 'The life and death of an aircraft. A network analysis of technical change', in W. Bijker and J. Law (eds) *Shaping Technology, Building Society*, Cambridge, MA: MIT Press.

Law, J. and. Hassard, J. (1999) *Actor Network Theory and After*, Oxford: Blackwell.

Midler, C. (1986) 'Logique de la mode manageriale', *Annales des Mines Gerer et Comprendre.*

Midler, C. (1993) *L'auto qui n'existait pas*, Paris: Intereditions.

Noble, D. (1979) 'Social choice in machine design: the case of automatically controlled machine tools', in A. Zimbalist (ed.) *Case Studies on the Labor Process*, New York: Monthly Review Press.

Ortmann, G. (1990) *Computer und Macht in Organisationen*, Opladen: Westdeutscher Verlag.

Orlikowski, W. (1992) 'The duality of technology: rethinking the concept of technology in organisations', *Organisation Science* 3, 3: 398–427.

Pettigrew, A. (1973) *The Politics of Organizational Decision Making*, London: Tavistock.

Pettigrew, A. (1985) *The Awakening Giant*, Oxford: Blackwell.

Pfeffer, J. (1981) *Power in Organizations*, Marshfield, MA: Pittman Publishing.

Swan, J. (1997) 'Professional associations as agencies in the diffusion and shaping of CAPM', in R. Williams and C. Clausen (eds) *The Social Shaping of Computer-aided Manufacturing and Computer-integrated Manufacture, COST A4 Social Sciences*, Luxembourg: European Commission.

Weick, K. (1990) 'Technology as equivoque', in P. Goodman and L. Sproull (eds) *Technology and Organisations*, San Francisco, CA: Jossey Bass.

Wenger, E. (1998) *Communities of Practice*, Cambridge: Cambridge University Press.

Williams, R. and Edge, D. (1996) *The Social Shaping of Technology*, Research Policy, 25, pp. 865–899.

5 'Push people's balls and push people's balls and push people's balls until something comes out'

Understanding implementation as a 'configurational practice'

Richard Badham, Ian McLoughlin and Karin Garrety

Introduction

Implementation is an act of collective will as actors determine what to do and mobilise resources to get it done. The masculinist quote used in the title of this chapter is one expression of the lived experience of implementation – the stress and frustration as projects get bogged down in what has recently been described by the head of Consignia UK as the 'treacle that just plays the game and gets in the way' (*Daily Telegraph* 2002). This chapter attempts to give some insight into the technical as well as organisational dynamics of this process, by drawing on and illustrating a 'configurational practice model' that integrates insights from 'implementation' and 'change management' approaches to change and innovation. This model is illustrated and elaborated through its application to a case study of work redesign in a traditional factory environment – a coal and coke plant in an Australian steelworks.

Configurational practice model

The configurational practice model has been developed to help conceptualise and explain the dynamics of complex socio-technical change initiatives such as total quality management and lean production, cellular manufacturing and group-based production, business process re-engineering and value-added manufacturing, socio-technical work redesign, high performance organisational redesign and so on. It emerged out of an attempt to understand the nature and impact of cross-functional technology initiatives such as the introduction of computer-aided design and manufacturing (CAD/CAM) (Badham and Wilson 1993; Badham 1993a). The discovery of the complex dynamics of local configurational processes in shaping CAD/CAM and its impact was found to be equally applicable to the introduction of self-managing work teams and cellular production (Badham *et al.* 1998). In seeking to understand these dynamics, insights from new approaches to implementation and change management were either helpful or confirmed the analyses being undertaken (McLoughlin *et al.* 2000). In contrast to much of the literature on change management and implementation, however, the configurational practice model focuses on the interdependence and integration of technological and organisational innovations, and seeks to draw on both research traditions in providing a more adequate explanation of the dynamics of local configurational processes and the nature and impact of initiatives to intervene in these processes to achieve either business or socio-political goals.

Implementation as reinvention

In reaction against linear models of innovation, diffusion and implementation is now widely recognised to be an active process (Figure 5.1). It involves processes of selecting, adapting and creating favourable conditions for the development and use of new technology – the effectiveness of which will be central in determining its impact and success (Badham 1993a). A recognition of such processes has led to a number of appeals to pay more attention to the implementation component of the innovation process (Braun 1985; Voss 1988).

Rogers (1995), drawing on the work of Charters and Pellegrin (1972) on educational innovations, uses the term 'reinvention' to capture the active nature of implementation – referring to

> the degree to which an innovation is changed or modified by a user in the process of its adoption and implementation . . . potential adopters on many occasions are active participants in the adoption and diffusion process, struggling to give their own unique meaning to the innovation as it is applied in their local context. Adoption of an innovation is thus a process of social construction.
>
> (Rogers 1995: 174 and 179)

Fleck characterises the dynamics involved in terms of the 'implementation equation': 'Successful implementation *requires* generic technology knowledge + local practical knowledge' (Fleck 1999: 246). Drawing on research on the social construction and shaping of technology, Fleck (1999) refers to this implementation process as one of 'innofusion', with the scope for reinvention depending on the degree to which technologies are well-established 'generic systems' or more flexible 'configurations' of more loosely

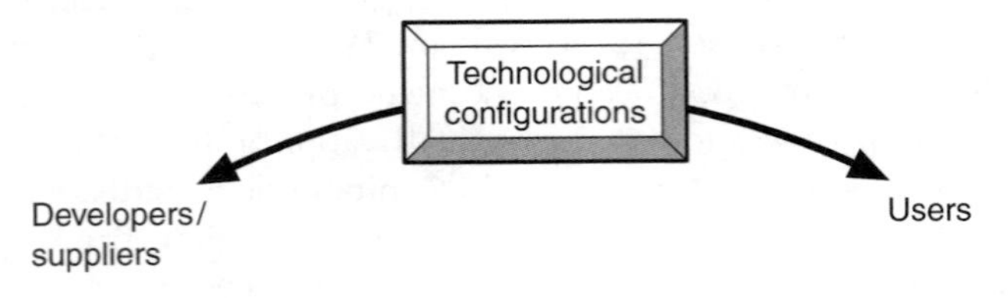

Figure 5.1 Implementation as reinvention

coupled and less well-defined elements. Fleck (1999) and Badham (1991) and Badham and Wilson (1993) observe that a number of technologies such as CAPM, CAD/CAM and robotics are more configurable than is commonly recognised, requiring adaptation and customisation in order to work effectively in different production contexts.

Fleck (1999) argues that the difficulty of synthesising local and generic knowledge means that the degree of reinvention required makes configuration type technologies far more difficult to implement successfully – a factor particularly notable in large firms introducing one-off innovations or smaller firms with little generic expertise in the innovations they introduce.

Rogers (1995) outlines a broader range of factors affecting the degree to which configuration or reinvention occurs, noting the importance of the degree to which:

- the technology is complex and difficult to understand;
- the adopters have knowledge about the innovation;
- the innovation is an abstract concept or tool with many possible applications;
- the innovation is a loosely bundled set of elements that are not highly interrelated;
- the technology is introduced to solve a wide range of user problems;
- there is a substantial degree of local pride in the innovation deeming it necessary for changes to be made to appear as a local innovation;
- an approach to change is adopted that encourages clients to become 'involved' in modifying the innovation.

Leonard-Barton (1995), drawing on a study of thirty-four projects that developed software tools to enhance internal productivity in four large US-based electronics firms, extends the analysis of implementation from a focus on the degree of reinvention of the innovation, to the more general process of user involvement and 'mutual adaptation' between the innovation and the organisational context into which it is being introduced. Mutual adaptation involves the 'reinvention of the technology to conform to the work environment and the simultaneous adaptation of the organisation to use the new technical system' (Leonard-Barton 1995: 104). This process is an iterative one, occurring in small and large recursive spirals, especially when initial smaller ones fail to work.

A key explanatory focus in this implementation research has been the developer–user relationship and its influence on the reinvention, configuration and mutual adaptation process. The logic of the reinvention process is often seen as a struggle between developers and designers trying to create 'reinvention proofing' in order to sell equipment, maintain quality control and speed up implementation, while adopters often see as desirable, and emphasise or overemphasise, reinvention in order to reduce mistakes and encourage customisation to fit with local situations and changing conditions

(Rogers 1995). This logic occurs at a number of different levels: in relationships between suppliers of equipment and user firms, as well as within user firms, between engineering and production departments, and information system departments and uses of information technology. Fleck (1999) emphasises the problems created for the effective implementation of configurational technologies as the result of failures in communication, economic separation, and the asymmetric distribution of knowledge and power between suppliers and users.

For those emphasising the active nature of the implementation process, the traditional neglect of the active role of the user is replaced by an emphasis on the key importance of users in processes of 'learning by doing' and 'learning by trying',[1] the significance of co-ordinating information and feedback between users and developers/suppliers and in the development process in general, importance of facilitating mobility of personnel and creating avenues for negotiation and formation of standards in industry sector user–supplier relations, the key role of particular users playing the role of 'translator' or 'mediator' (Trigg and Bodker 1994; Okamura *et al.* 1998) in organisations, and the broad range of users involved in any implementation context (Suchman 1999).

This has led to the creation of a variety of prescriptive remedies for overcoming what is seen as a neglect of users through the dominance of a technocratic/developer logic (Perrow 1983) and encouraging user involvement and participation. Leonard-Barton (1995) emphasises the importance of user involvement (particularly what she calls 'co-development' or 'apprenticeship') and forced knowledge sharing, especially important when the developer does not have local knowledge. Given the difficulties involved in configuring technologies to local contexts, these processes are inherently uncertain, as Fleck (1999: 256) emphasises: 'in the case of the development of configurational technology, industry is the laboratory.'

While this approach has opened up the intellectual space necessary to explore the configurational dynamics of the implementation process, there has been a tendency to restrict the focus to the reshaping of equipment and its requirements for use, and developer–user interactions in this process. Less attention has been paid to the implementation of organisational techniques, the configuration of the organisational as well as the technical aspects of production systems, and the complex relationships between different types of more or less active users within the adopting context. Suchman (1999: 258), for example, argues that it is necessary to abandon 'the myth of the lone (male) creator of new technology on the one hand, and the passive recipients of new technology on the other', and to broaden analysis beyond a focus on simple designer/user conflicts. There is a need to move 'from a view of design as the creation of discrete devices, or even networks of devices, to a view of systems development as entry into the networks of working relations – including both contests and alliances – that make technical systems possible' (Suchman 1999: 258). Simple ideas of designers and users are inadequate, for 'just as the term "designer" opens out, on closer inspection, onto an extended field of alliances and contests, so does the term "user". Organisations comprise multiple constituencies, each with their own professional identities and views of the others.' This view is confirmed in studies of computer system development and use by Trigg and Bodker (1994). They conclude that

> In communities of every day users of technology the blurred designer/end-user distinction is the basis for a new community of practice, people who behave like designers and as well as users, and in whose hands the success and long-term survival of an installed technology often rests. Researchers studying the phenomenon use labels like tinkerer, translator, and gardener.
>
> (Trigg and Bodker 1994: 45)

Technology use inevitably involves recontextualising technologies that are designed at a greater or lesser distance from the local site, and in this process there needs to be a greater extent of humility about the extent of any actor to control technology production/use, and the importance of valuing partial translation, partial integration and respect for heterogeneity rather than simple developer or user models of homogeneity and domination (Suchman 1999: 264).

Management of planned and emergent change

Within the field of research on change management, there has traditionally been a focus on planned organisational change rather than the implementation of equipment. While the classification of types of change is often broadly defined (Badham 2000), most attention has traditionally been paid to planned changes in organisational redesign, interpersonal relations and corporate culture. Similar to the developer/user focus of implementation studies, change management research has traditionally tended to focus on the active nature of leaders and change agents on the one hand, and resisters on the other. In analysing the dynamics of planned organisational change, there has been a central preoccupation with the dynamics of leadership and change agency – the former focusing on leader/follower dynamics and the latter on the skills, techniques and roles of designated change agents. In contrast to implementation studies, however, the main focus has been on relationships within the organisation undergoing change rather than inter-organisational relations between suppliers and users of change methods and techniques.

In reaction against an overemphasis on the character, actions and power of leaders, there has been a trend towards analysing and understanding the key role of 'followers', the existence of leaders at all levels of the organisation, and the supportive 'post-heroic' nature of successful leadership styles (see Figure 5.2) (Clegg and Hardy 1997; Fulop and Linstead 1999). Similarly,

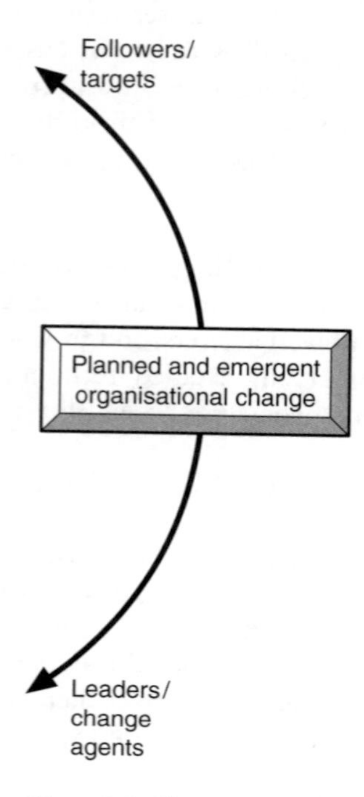

Figure 5.2 Change as improvisation and emergence

the exaggerated stress on the role of formally designated change agents has been replaced by more interest in and concern with the 'complex cast of characters' involved in building effective coalitions for change (Buchanan and Storey 1997; Buchanan and Badham 1999). Others have introduced the importance of the strategic impact on change of broader contextual and inter-organisational factors (Dunphy and Stace 1990), yet have been criticised for their neglect of the internal micropolitics and emergent nature of change processes (Orlikowski and Hofman 1997).

The more pluralistic or processual approaches to change also provide a closer examination of the perspectives, motivations and effects of the different sub-units, departments or groups involved in the change process. Bacharach, Bamberger and Sonnenstuhl (1996), for example, explore the degree of 'cognitive dissonance' between the 'logics of action' of actors at the institutional, managerial and technical levels of organisations, and the way in which these become, or fail to become, aligned during the change process. Yet despite the increasing sophistication of such models of change, there has continued to be a relative neglect of the complex dynamics shaping the technical content of change and its outcomes for the change process (McLoughlin *et al.* 2000). In recent years, there has been growing interest in the production and consumption of 'management fads and fashions' (Benders and van Veen 2001), but this has not yet developed into a solid body of research on the dynamics of supply and use of such management techniques. More recently, there has also been a greater concern with understanding and informing agency and improvisation in change processes (Orlikowski 1996; Buchanan and Badham 1999). Orlikowski (1996: 65) points to the importance of incorporating what she calls 'emergent change', i.e. 'the realisation of a new pattern of organising in the absence of explicit, a priori intentions. Such emergent change is only realised in action and cannot be anticipated or planned.' Her focus on this type of change derives from what she calls a 'situated change' perspective, which sees organisational transformation as 'an ongoing improvisation enacted by organisational actors trying to make sense of and act coherently in the world.' (Orlikowski 1996: 65).

Unlike implementation studies, organisational change research has adopted a broader focus of study beyond the narrow confines of equipment and technical systems. It has also provided more insight into the influence of the complex relationships and interactions between different actors within the organisation undergoing the change. Unlike implementation studies, however, there has been a relative neglect of the technical content of change, and less attention paid to the inter-organisational dynamics between suppliers and users of organisational change methods and techniques.

Configurational practice

The configurational practice model seeks to build on and assist in integrating the implementation as reinvention and planned and emergent organisational change approaches to innovation, as schematically outlined in Figure 5.3.

Similar to the implementation as reinvention approach, the configurational practice model takes the technical content of change seriously, and focuses attention on the shaping of this content by a complex array of developers and users. Drawing on the management of planned and emergent change, however, the focus of the configurational practice model is on organisational as well as technical innovation, and the developing interactions between those more highly active in promoting the change and those whose participation is required yet play a less central role in initiating the change process. In accordance with recent research on both implementation and change management, however, the model of practice involved stresses the uncertain and emergent nature of local configurational processes that makes them less reducible to rule-governed behaviour or amenable to prescriptive methods. It is also an important feature of this model that the production system into which a new technique or change is being introduced is itself a configurational process, and that implementation is a matter of interaction between the introduced 'technique' and the ongoing dynamics of the production system understood as a configurational process.

The configurational practice model outlined in Figure 5.4 and presented

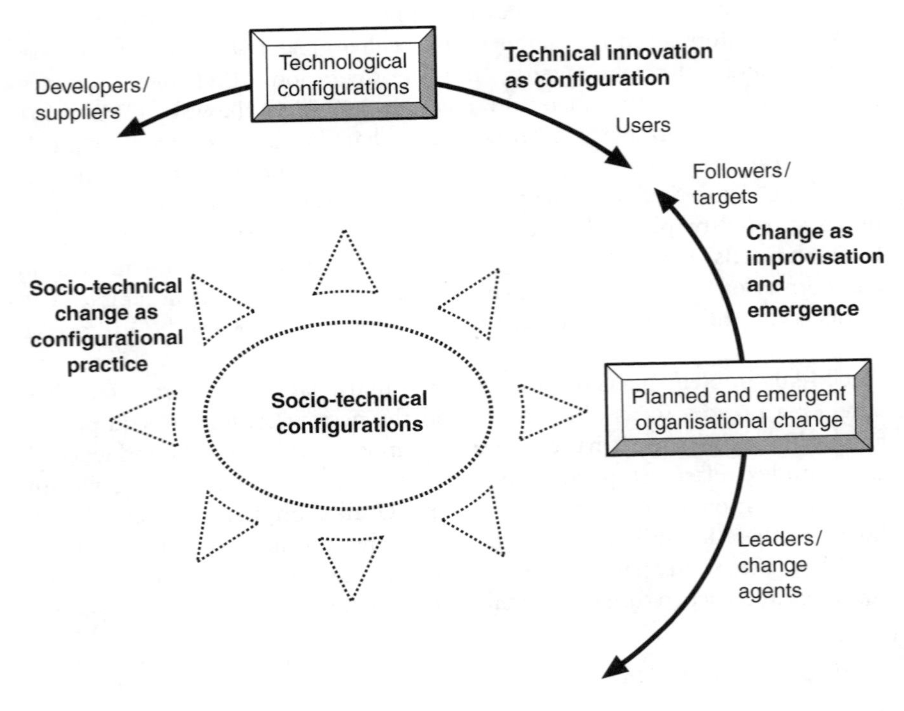

Figure 5.3 Configurational practice: socio-technical implementation and change

in more detail elsewhere (Badham *et al.* 1997), is based on six main assumptions about the implementation process and the 'system' undergoing change.

It is local

As outlined in the reinvention literature, generic innovations are, and have to be, actively configured to *local* contexts in order to be workable. This is true both for the technique being implemented and the working of the operational environment into which the technique is being introduced. While the degree of local freedom and configuration varies, and may be more or less constrained by circumstances, it is an often inadequately recognised and essential element of all implementation processes.

It is interdependent

Extending Leonard-Barton's (1995) model of 'mutual adaptation', and in line with the socio-technical systems framework (Badham 2000), whether the innovation being introduced is material or social, technological or organisational, it will involve, and have repercussions upon, both technical and organisational dimensions of local production contexts. This

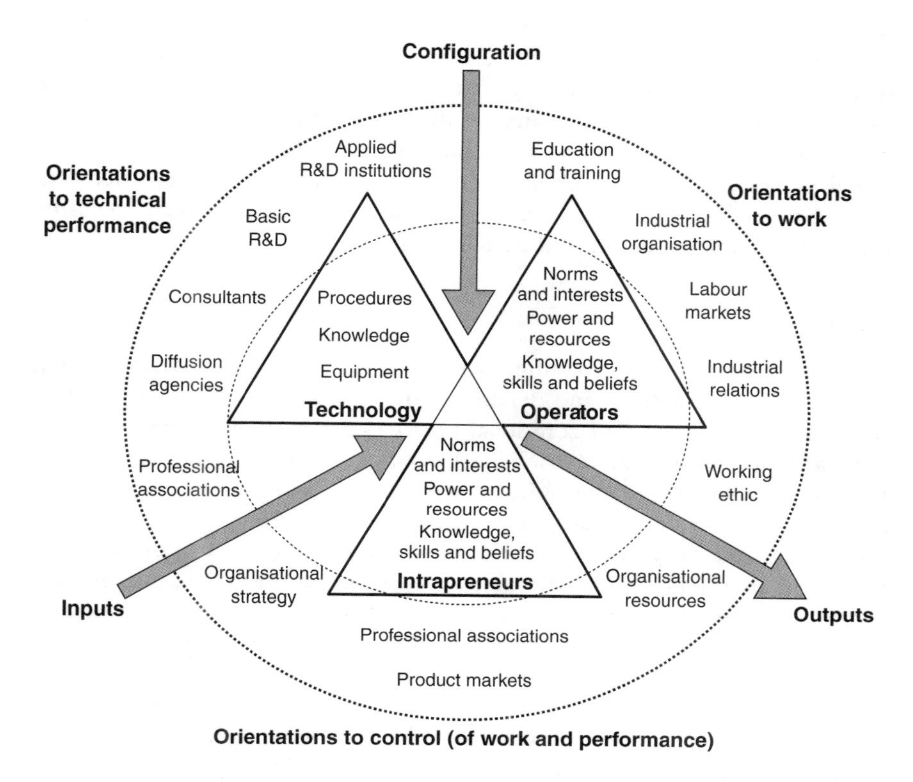

Figure 5.4 Configurational practice model

interdependence is recognised in research ranging from ergonomic and work psychology studies of methods for the 'allocation of functions' between people and technology to sociology of technology studies of the social character of interpretations of the nature, working and effects of technology and the 'heterogeneous engineering' of networks of interchangeable technical and organisation 'actors' (Badham 1992; MacKenzie and Wajcman 1999).

It is complex

Local and interdependent configurational practices constitute a *complex* interactive process. This complexity has rarely been captured by simple views of relations between the 'technical' and the 'social' (Einjatten 1993). On the one hand, both technical and social components are mutually configured in a dynamic process of local customisation and change. On the other hand, the very use of the categories 'technical' and 'social' has often failed to capture the complex organisational dynamics involved in the 'social'. Research on implementation and change has focused our attention on developer/user and leader-change agent/follower–client or target dynamics, and studies of the social shaping of technology have detailed the key importance of 'heterogeneous engineers' and socio-technical 'system builders' in creating effective socio-technical networks. Drawing on this work, the configurational practice model emphasises the importance of the nature and interactions between production system 'operators' and 'intrapreneurs' within local configurational processes. This interaction is complex, however, since operational or intrapreneurial *roles* in a configurational process may be played by the same actors or by different actors at different times.

It is diverse

The different elements of configuration practices are *diverse*, with varying linkages, goals, emotions and dynamics. In contrast to systems approaches, it is not assumed that configurational practices have a single purpose or that innovation or reproduction in ongoing configurational processes is assured. In contrast to views of the object under change being a technical, operational or production system, the focus is on more loosely integrated and potentially mutually contradictory configurational processes. In addition, the 'open' nature of the configurational processes means that there is a continuing interaction between the broader 'macro' context of the process under consideration and its 'micro' elements, and these interactions frequently impose constraints or 'pull' elements of the configurations in different directions. This interaction between the 'local' and 'broader' contexts and the problems it creates for actors involved in configurational processes is an ongoing source of tension and difficulty (Law 1997).

It is emergent

The complex interactions within configurational processes, and between these processes and their context, mean that the dynamics of change are inherently unpredictable and unable to be reduced to simple rules or read off as the result of closed system operations. Jervis (1997) emphasises that such complex systems are characterised by the fact that: results cannot be predicted from the separate actions of elements within the system; the outcome of strategies of actors within the system depends on the strategies adopted by others; the behaviour of those within the system changes the environment that then acts back on those behaviours and their effects; and the result of the first three is that outcomes are often not intended, and are better seen as unintentional consequences of actions.

It involves practice

Each of the configurational components are combined in situated *practices*, where action and agency is constrained and enabled by the conditions in which it occurs. These practices are where the 'rubber meets the road'. They are conscious and tacit, planned and emergent, stable or patterned and novel or changeable, the result of action and the product of circumstance. The exercise of will, and reflection on strategies and conditions of activity, is an essential component of agency and needs to be recognised as such if individuals are to act to change their circumstances. However, the outcomes of this action are constrained by both known and unknown conditions, and are likely to vary from the intentions of their 'authors'.

In the following three sections this view of implementation will be illustrated and used in outlining a case study of the initial nine months of a work redesign project in a coke-making plant attached to a large steelworks in Australia. The first section outlines the general configurational process involved in the introduction of a particular work redesign method into this specific site. The next two sections focus on two specific issues, among many, raised in the work redesign process, using these issues to explore in more detail the nature of the configurational process.

Case study: work redesign in a coke-making plant

This project is taking place in a coke-making factory in Australia, employing approximately 480 employees in supplying five million tonnes of coke per annum to its main customer – a blast furnace.

Work redesign: the onion model

The work redesign initiative is directly targeted at both the operator and intrapreneur components of the plant's configurational process that transforms incoming coal from local coalmines and mines in Queensland into coke which is then transported to fuel the nearby steel furnace. In line with traditional socio-technical work redesign models, the 'technique' being introduced has both a process and content component. Traditional socio-technical design adopts an open socio-technical systems method of analysis and design frequently accompanied by a more or less explicit advocacy of job enrichment and semi-autonomous work groups as a work design solution (Badham 2000).

Within the socio-technical model of the process to be followed, different proponents advocate more or less expert-led or participatory design processes. The model of the redesign process being employed in this project has been developed by a consultant with substantial experience in both socio-technical systems design and interpersonal organisational development. His experience with traditional expert-led socio-technical redesign has prompted him to advocate a more strongly 'participatory design' approach, paying considerable attention to extensive workforce involvement and the establishment of supportive interpersonal relationships in the work redesign team as well as within the work system being redesigned. In this project, the work redesign team is characterised as:

- A non-decision-making body (i.e. the conclusions that it comes to are not binding on management or the workforce, and it does not get involved in negotiations over job demarcations, numbers and wages).
- A non-representative body (i.e. the individuals on the work design team do not represent specific groups or constituencies. They are part of the group as a result of the knowledge and experience they can contribute).
- A model for cultural change (i.e. while the formal objective of the group is to come up with one or more work redesigns, the way in which it carries out its work is intended to be a model of the way in which group work is to be carried out in the future).
- An open forum (i.e. one in which interested visitors are allowed and encouraged to attend at any time).

In contrast to many other work redesigns, there is no steering committee to which the work design team reports – the outcome is intended to be the facilitation of broader workforce discussion of possible work redesign and the development of proposals to assist this discussion. In order to assist the discussion process, however, the consultant has produced a 'road map' or 'process' to be gone through: the 'onion model' (as nicknamed by the work redesign group: see Figure 5.5).

As illustrated in Figure 5.5, the process begins with the 'core' activities. These involve:

- establishing a vision and mission for the work redesign group, and establishing the values they will work by;
- setting the strategic objectives for the work redesign project;
- identifying the key stakeholders and their interests.

Once these are established, the work redesign group can begin with its basic technical and social analysis, involving:

- layout of equipment;
- nature of technology – in particular stages of the process and their inter-dependence;
- establishing job design principles and creating new job definitions;
- establishing team-work principles and the nature and scope of work teams.

This 'inner ring' is followed by identifying, investigating and designing the support systems for the new jobs and teams, including such areas as reward systems, performance management, information systems, training and so on. At each stage of the process, the consultant recommends feedback to the general workforce and collection of information and opinions before proceeding to the next stage.

At times during the process, the consultant has revealed his ideas about

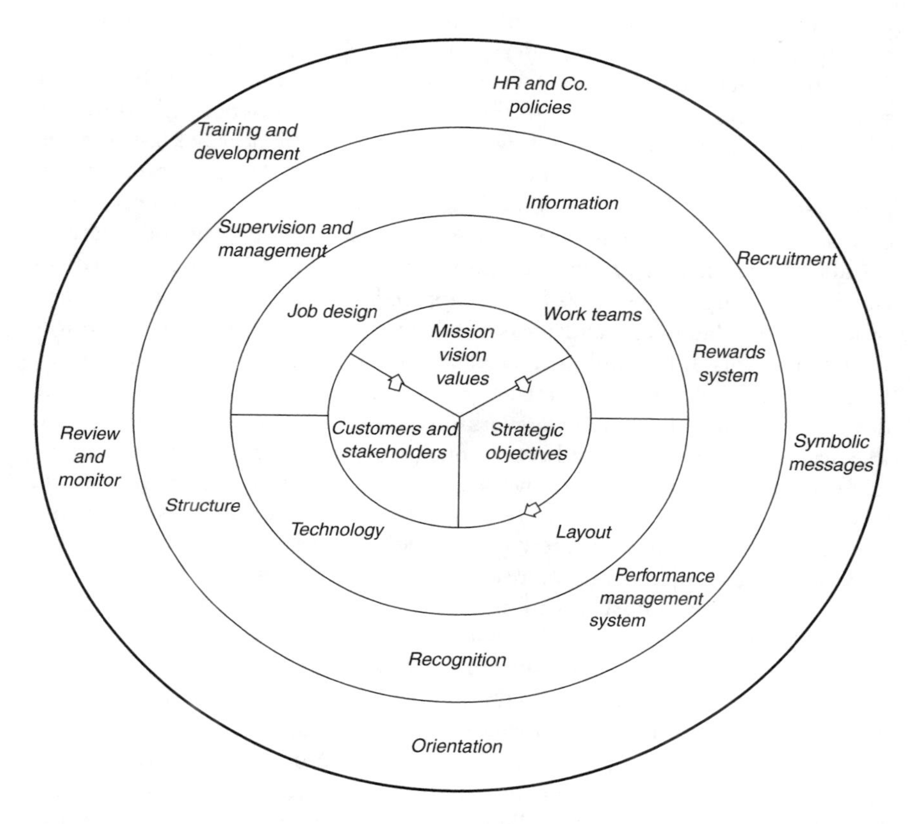

Figure 5.5 The 'onion model'

the end solution of the work redesign. He has presented a strong general view of the transition from a 'rule-governed' workplace to a 'values- or purpose-driven' workplace, a particular view of production as dominated by semi-autonomous 'manufacturing teams' with 'support' from 'flying teams' of 'coaches' and indirect groups such as mechanical and electrical maintenance and a managerial 'strategy team', and an outline of values and principles for governing behaviour in the new workplace (e.g. honesty, caring, commitment to fulfilling stakeholder objectives).

This model of work redesign has emerged from a process of tailoring the consultant's earlier method to the configurational process within the plant. Initially, the plant manager and the human resource officer in the plant hired the consultant to carry out a form of work design different to past practice within the company and to that carried out by the consultant, and this influenced the proposed process that the consultant ultimately put forward to the plant. Over the past eight years, work redesign in the company has often been carried out by a cross-sectional slice of the workforce who are assigned to the project full-time, and spend nearly six months working on the new redesign facilitated by an external consultant. This group then presents its recommendations to the workforce which, after some discussion, takes a vote. On the basis of their experience in leading one of the most radical and successful work redesigns in the nearby blast furnace, the plant manager (PM) and human resource officer (HRO) are critical of three aspects of this process. First, the time taken for implementation is excessively long, as 'locking away' a design group for a long period created a distance between this group and the workforce. Worker representatives in the design group were seen as being 'captured' by management, and the gap of knowledge, language, and ideas between the design team and the workforce led to misunderstandings, suspicion and conflict. The result is often a long and difficult implementation period. Second, the distancing of the design team from the rest of the workforce contributed to a now widely held idea that the work design had been 'done', and that the new system was now fixed. Rather than viewing the work design as an initial and partial design, or one that has to be adapted to new circumstances, the design was seen as a one-off expert solution, with few in the workforce having the skills, inclination or interest to continue or resurrect work design activities. Third, the work design is often pushed through with the help of a supportive coalition of middle managers, unionists and shop-floor operatives. The assumption is often made that once the new system is implemented, others who have previously resisted the change would gradually 'come on board'. However, when the PM left the blast furnace after the redesign, no more change seemed to be occurring, and in his words the 'process has stalled. It had become too dependent on me. By communicating directly with the shop-floor, I had bypassed many middle managers, whose support is essential over the long haul.' Consequently, the HRO and the consultant prepared an alternative work design process. The work redesign group (the 'working party') is not full-time. It meets for one day each week with additional time taken for special activities (e.g. communication, visits, training) when necessary. The external consultant, rather than facilitating the process full-time for four months, returns infrequently to help the group overcome any problems it is having and to check on the process. This veers strongly from the way the consultant has worked in the past, and even what was intended at the start of the project.

The absence of the consultant as a full-time facilitator increased the degree of uncertainty and misunderstanding about the nature of activities to be conducted at different stages of the 'onion' process. With regard to the 'technical' analysis, for example, not only did the consultant provide a 'technical matrix' model that differed from the classical technical 'variance chart' but the analysis was variously seen by the work design team as an investigation of interdependencies in the changing material process, the nature and problems of equipment operation at different stages of the process, 'variances' or problems at different stages, and an investigation of problems of equipment and people at different stages of the process. The nature of job design and team work was variously interpreted to refer to drawing up some general principles of good work and team work, to creating a 'responsibility matrix' outlining the changing tasks to be taken on by teams, to a detailed specification for the nature of jobs and group work in the future. As the group became more uncertain and frustrated about the activities to be conducted at each stage, they began to question the past successes and credibility of the consultant as well as the nature of the 'onion' process. While the process came to be seen as merely a 'road map' rather than a solution, the group returned at various stages to a faith that if only they 'followed the process', a favourable outcome was likely to be achieved. Others expressed real confusion: 'If we are following the onion model, and a lot of us don't know whether we are or not, are we now looking at work teams, or are we drawing up scenarios which include all the rings in the onion model?' In the absence of such certainty, the group turned increasingly to the university to fill in some of the details from their past experience and expertise in job design and to help overcome their self-proclaimed 'barriers' to proceeding with the job redesign. Following a brief introduction to work redesign in a three-day workshop at the university, the group began to focus on making explicit their 'implicit contracts' threatened by job redesign, seeking to identify the 'bucket of tasks' to be carried out by the team, and began preparing and presenting alternative work redesign 'scenarios' to the workforce. In addition, the HRO began to exert more influence over the content of discussions, and became more critical of the excessive 'ownership' shown by the consultant and the university researchers to the perceived relative neglect of the independence and autonomy of the group. The PM began to weaken the emphasis on formal job design, stressing the original intention as being one of modelling a new culture and form of behaviour, with a reduced focus on the importance of coming up with a design that was effective and acceptable to the workforce.

The dynamics of the introduced work redesign process was influenced by a number of factors other than the structure and processes outlined by the consultant in the onion model. The company had previously introduced 'rational' Kepner-Tregoe processes for 'problem-solving and decision-making' in teams. The company had also carried out highly emotive interpersonal relationship training for all supervisors and superintendents, and the general principles underlying this training were also introduced into this work redesign process by those who had been through this training and by the consultant who had provided some of this training in another context. In addition, the nature of facilitation occurring in the group had an ongoing changing character influenced by: the time available from the consultant and the desire of the HRO and the group to have him attend; the preferences of the HRO responsible for the project and her absence from the plant for some time; the ideas and preferences of the superintendents who at different times felt more or less desire to control the process; the increasing degree of facilitation ideas and interventions from all work redesign team members; and the uncertain and changing role of the university personnel who acted at different times as observers, evaluators and providers of process and content ideas.

With regard to the content, the proposal put forward by the consultant for a general 'manufacturing team' for the batteries was not a logical derivative of the commitment to a 'value-based' organisational structure. The PM stated in a personal interview that the consultant's model was 'my model', and his view of a manufacturing team was strongly influenced by the conclusions of a divisional level plant manager's meeting held midway through the work redesign process, where the general character of 'manufacturing teams' was outlined. The view of manufacturing teams outlined by this divisional management team was highly general and flexible, leading to uncertainty about what was implied. The PM, for example, interpreted it to mean that maintenance personnel could be in the manufacturing team. Other plant managers, however, saw maintenance as being subcontracted, and there was a tone in the outlined 'vision' of the manufacturing team that emphasised reduction in numbers, potential subcontracting of maintenance services to the production teams and/or the teams taking on routine maintenance, and not employing people permanently for 'peak' production requirements and only for 'normal' operations.

This process of interpretation and reinvention may be attributed to the existence of some ambiguity about the content of the 'technique' being proposed to change the configurational process, or a gradual reinterpretation or 'filling in' of a general vision provided by the 'technique'. In part, the ideas of the consultant about the final nature of the proposed redesign were deliberately left at a very general level and not imposed upon the group. This approach was undertaken in direct contrast to earlier work reorganisation initiatives in the company. A colleague of the consultant, who had undertaken an earlier redesign in a nearby section of the plant, even commented on this earlier initiative that 'it was not really a job design at all, merely manipulation by the consultant to get the group to accept his prior solution'. The lack of imposition of a 'content' solution by the consultant was influenced by both the desire to get the group to develop their own designs in consultation with the workforce and the less than expected degree of facilitation by the consultant. It has also led to a greater push for the introduction of a total productive maintenance (TPM) manufacturing philosophy promoted by an alternative Japanese consultant as a content guide. When, midway through the design, the work redesign group was crudely informed about the details of the divisional management 'manufacturing team' approach, and particularly its strong focus on the subcontracting of maintenance, there was a strong response about 'hidden agendas' of management that underlay the explicit consultancy method and technique with which they had been presented. This suspicion reflected a continuing suspicion by shop-floor employees that the method and technique presented to them was not simply a process with an open ending but was a cover for quite specific content proposals that were in the mind of management, and that management was attempting to get them to 'feel involved' and, if they came up with proposals that differed from management's view, they would be reversed.

The configurational process within the plant

Technology

The plant within which the onion model process was introduced has three sections: coal preparation or 'washeries' (where coal coming in is washed and ground); the 'batteries' (where the ground coal is heated and re-formed into coke), and 'gas processing' (which extracts the gas from the process and turns it into gas products for sale inside and outside the larger plant in which the coke-making factory is located). The nature of the core coke-making process was clearly defined by one superintendent as 'to fill the ovens and push them on time', i.e. a cycle of filling or charging, heating, and pushing out coke from the ovens or batteries – where ovens get 'pushed' approximately every twenty minutes. The plant emits a significant amount of pollution, even though this has been reduced ('you used not to be able to see down Battery Road because of the smog!'). It is still a dirty place to work and dust in eyes is a common injury. For those working outside on the plant, or on the machinery in the batteries, the environmental conditions can be both hazardous and unpleasant, getting very hot in the summer. Whatever the process of work redesign, the traditional structure of the batteries imposes certain constraints. Given the existing level of automation, many operators are tied to machines at jobs that are geographically separate from each other, a clear restriction on substantial levels of effective in-process team work. In addition, a number of initiatives to increase automation are premised on 'operatorless machines' that leave little room for enriched 'super-operator' system intervention and monitoring jobs.

Operators

The plant is unionised with three unions represented on site. Despite increased job rotation brought about by a job reduction ('restructuring') initiative in the early 1990s, jobs remain fairly narrowly defined, with little job enlargement or enrichment, and many operations jobs tied to the equipment responsible for 'pushing ovens'. The culture of the plant is one of low self-esteem (it is commonly seen as 'the arse end' of the company's operations and there is a policy of no forced transfer to the plant), and a traditional distrust of management and resentment of their privileges (although no longer observed, there is still a 'staff' and 'wages' toilet on the batteries). The workforce is relatively highly paid and currently protected by a management/union steel agreement, yet the picture is not all rosy for the employees. The agreement is coming to an end, the industry is under such severe pressure that the whole plant may be sold off in the near future, there is a relatively high level of unemployment in the region, and many would find it difficult to get jobs elsewhere given the relatively unskilled or highly specialized coke-making nature of their jobs. Everyday work operations are characterised by a substantial degree of conflict, breakdowns, inefficiencies and different groups successfully exploiting their abilities to create 'free time' at work – one HRO commenting that 'a rule of thumb is that you get three to four hours' work out of an eight-hour day from the workforce'. As with many such process operations, there is a substantial division between different groups of operators who are tightly coupled to different parts of the process (e.g. charger drivers, heating regulators, ram drivers), between semi-skilled output-oriented operators reporting to production supervisors and skilled support maintenance electricians and fitters, many of whom report at least in part to non-production supervisors, and operating midway between these two are utilities operators who maintain the batteries themselves. All these conditions have influenced the response of operators to the work redesign process. It is commonly viewed as merely another job-cutting restructuring exercise, participation is widely seen as a cover for the 'hidden agenda' of management, and there is little optimism or understanding of how jobs could be enriched given the traditional low level of skill and motivation of the workforce and the entrenched nature of demarcations. Changes in the external environment are, however, increasing the degree of uncertainty about the possible nature of change. In particular, changes introduced in the associated 'coal preparation' area of the plant and the attempted sub-contracting of maintenance by the company at a broader divisional level has fundamentally threatened the established and privileged position of maintenance workers, who are currently claiming to be highly amenable to being 'part of the manufacturing team' and devolve routine maintenance tasks to operators.

Management and intrapreneurs

The management team at the plant is headed up by a new and charismatic plant manager supported by four line superintendents (one from each of the main operational units in the plant), and three senior technical officers, one of whom has responsibility for external liaison. The plant manager also has one HRO and a maintenance superintendent reporting to both him and their functional heads outside the plant. Underneath this team, there are approximately sixty senior supervisors and supervisors in a workforce of 480. The individuals in this team have worked for most of their lives within the company's hierarchical and bureaucratic structure. Despite being introduced to highly emotive participatory leadership training, none (apart from the HRO) have been involved in driving real change projects designed to break down established structures and hierarchies. Despite closure of part of the plant the number of superintendents has not yet been reduced, and there is still a disproportionate number of supervisors. With a concern for their own jobs, and grappling with the changing rhetoric of culture change, neither the management team under the plant manager nor middle-level line management are highly skilled, experienced or confident agents of change. In addition, neither a weekly middle management 'operations forum' nor the work redesign meetings have led to an overall increase in committed and enthusiastic intrapreneurs for change from the middle and lower levels of the organisation. One pilot team has, however, addressed this imbalance to some extent, and there are increasing indications that members of the work redesign group are acting more as intrapreneurs – actively communicating and promoting both the need and basic substance of change to the broader workforce.

Key issues in the configurational process

It is neither within the scope of this chapter to explore further the overall process of reinventing or configuring the generic work redesign method, nor the influence of the overall configurational process on this reshaping activity or the effect of the work redesign on the general configurational process. What we shall do, therefore, is to illustrate the main issues by isolating for more detailed discussion two major issues in which the generic work redesign method has been configured in the process of its application in the project.

Representation and decision-making

All socio-technical redesign projects have some form of stakeholder representation and participation by those with the knowledge and expertise necessary to carry out the redesign. Different arrangements are created for involving those with the necessary knowledge in the design process and for reporting back to those stakeholders making the final decision about the work redesign to be adopted. As we saw above, no stakeholder committee was set up for this project, and the work redesign team was set up to be a non-decision-making body, a non-representative body, a model for cultural change, and an open forum. What was meant by non-representative was that individuals were selected for their knowledge and experience rather than for their representation of an interest group, and no decisions were to be made by this body of 'representatives' that would be binding on outside constituencies. As the consultant put it,

Dave: I don't want you to have the burden of representation. You are here because of your individual experience and view of the world and what you bring to the group. You do not represent a constituency. And you are not a delegate, that is the role of consultative committees and trade unions.

This was explicitly set up in order to avoid the development of a conflictual 'industrial relations'-type environment, and to encourage the different participants to contribute their knowledge towards the development of solutions in the interests of the overall business and workforce. With members drawn from superintendents (four), production and maintenance supervisors (three), maintenance fitters and electricians (three), HR (two), and operators (five), the work redesign team covers a broad spectrum of both staff and wages within the workforce. This concept has been subject to ongoing detailing, reinterpretation and renegotiation during the work redesign project.

The degree of 'local' configuration has been evident in a number of areas. For example, as there is no stakeholder representation, individuals from different areas found themselves in a difficult situation. They were encouraged to contribute knowledge and experience 'from their area' but are not expected to represent the 'interests' of themselves and their mates – a task which, not surprisingly, they found rather uncomfortable. This has resulted in a lot of frustration, distrust of the honesty of different groups and talk about the costs and benefits to the different groups outside the official group discussion process. Influenced by the university researchers, the different members have been encouraged to recognise their 'implicit contracts' and make them explicit as part of negotiations over the future design. A compromise position has informally developed whereby the design team is creating 'scenarios' for discussion by the workforce rather than a final solution, but that in this process they are able both to consider the broader interests of the business and the workforce and attempt to find out the views and interests of the groups and seek to integrate these into suggested scenarios.

The interdependence between the 'technique' and the 'social' is evident in the way in which the issue of 'representation' is interpreted or negotiated by participants with a traditional low level of trust between the groups within which they work. Issues come up, for example, in terms of the numbers of people who should participate from different sections of the plant. This has been raised prominently in discussions of why all three senior superintendents (Kim, Tom and Andy) are involved in the work redesign team. One interchange reveals this distrust within as well as between groups:

Dave: Kim is not really here as a superintendent.
Joe: Kim is here for being anally retentive.
General: How many superintendents should we have?
Tom: If Andy is on it, then I need to be on it, because I don't trust the bastard, I like Andy, but we think differently, and if I wasn't on it, I would find ways of stuffing it up, not deliberately, but ways in which we do this . . . originally I heard Kim was going to be on it, and then Andy and I wouldn't be on it, but if Andy was on it, I had to be.

Whether or not the individuals or groups are formally defined as 'non-representative', there is clearly a substantial degree of representation of individual and group interests.

The complexity of the configurational process quickly became apparent as attempts were made by the group to work out the way in which the non-representative work design group was integrated with other representative groupings in the workforce. An initial exercise in addressing 'cleaning' issues was criticised for doing the work of the occupational health and safety committee. A union delegate for the electricians attended one meeting to criticise members of the work design team for making statements that they did not have the authority to make, for example, with regard to electricians operating machines as part of their jobs. Midway through the work redesign process, the consultant criticised himself for not providing a mechanism for the work design team to communicate with the plant management team.

The work redesign process has been influenced by traditional modes of representation in the plant, the desires and interests of the members of the work design team, and the interests of the PM, the HRO, the consultant and others in achieving specific outcomes. Many times these have come into sharp conflict. The onion model process displays similarities and differences to other work redesign activities and philosophies used by the HRO, the university researchers and other HR officials who have attended the group. The work redesign process also draws on and in part conflicts with other process methods used by the members of the group. The members of the group are accustomed to different methods of communication (slanging, joking, conflict and recrimination), and much of the consultant's time was spent in attempting to influence these methods of communication. Some members of the group were also less ready to suspend and take criticism of the 'sacred cows' held by their group, leading to two or three members leaving the work design team and being replaced. Finally, the goals of the PM, the HRO and the superintendents have also conflicted with each other and the goals of the consultant – the PM content with the work redesign exercise stimulating discussion and energy within the workforce without coming up with a final and definitive work redesign proposal, the superintendents clearer about underlying production process outcomes they want from the redesign than the means of getting there, and the HRO at times in conflict with the consultant and the university personnel over the degree to which the work redesign team should be left to define its own process.

All these developments led to emergent changes in the mode of representation as the participants attempted to grapple with the problems they raised. The idea that the work redesign team was not a 'decision-making' body led to an unexpected inability of the group to put forward work redesign proposals, on the grounds that it was not its role to make a 'decision'. This reflected and reinforced the reluctance of many in the group to address and resolve fundamental conflicts and tensions between 'wages' and 'staff', 'operations' and 'maintenance', 'workers on different batteries', and so on.

When the university researchers suggested developing scenarios, the group latched on to this idea and pursued it with enthusiasm, somewhat to the chagrin of the consultant who did not see the presentation of scenarios as the result of the 'non-decision-making' process but, rather, the creation of one proposal. In addition, the idea of avoiding viewing participants as representatives, and the kind of 'voting' system associated with this, led to a lack of understanding of how to resolve disagreements in the group. As illustrated in the following interchange, the group is still caught up in trying to define what constitutes agreement or consensus in order to move on after a discussion.

Dave: This is not a forum where majority rules, brainstorming activities are a focus for achieving understanding and consensus.
Other: Is there a focus on full agreement?
Dave: No – consensus means if I don't 100 per cent agree with it but I can live with it.
Other: Are there any actions taken against people who don't conform to these things?
General: Discussion about actions taken against people who do not conform.
Other: Working party is not a judge and jury . . . what is the difference between agreement and acceptance?
Craig: I often accept what my wife says, but I hardly ever agree with it!

Joint optimisation, caring and tar

Socio-technical work redesign recommends the joint optimisation of the 'technical' and 'social' components of work systems. The social component is often taken to refer to some combination of ideals of a good job and quality of working life considerations. One issue that has always dogged socio-technical redesign is the degree to which ideals of job enrichment and semi-autonomous team work, often held by socio-technical consultants and researchers, are actually held by the workforce being asked to change, and, where there is a conflict, how this is to be addressed. During the late 1990s, as work redesign often occurred with restructuring and lay-offs, a broader issue became more widely considered: the importance of 'caring' for the workforce during the change process, and ensuring 'employability' rather than simply 'employment' given the decline in job security. Joint optimisation often comes to take on a different meaning, as caring and employability become intertwined with more traditional qualities of work-life considerations.

During the work redesign in the case study plant, the issue of caring for the workforce has come up in a number of different contexts, particularly in the form of the degree to which the interests of specific groups will be 'traded off' during the redesign. One particular issue of interest emerged, quite unpredicted, surrounding the deposits of tar that sometimes occurred on employees' cars in the local plant car-park. The general issue of caring became transformed into an issue of how to deal with the costs of having cars detailed after getting tar on them (and an interlinked issue of responsibility for paying for car windscreens that cracked in the plant carwash installed to clean pollution off employees' cars). Some of the issues are raised in the following two quotations.

Craig: My car cost me $23,000, more money than I have ever had, and I can't afford another like you. And it is being destroyed by this place, and you don't care. Now people say cool down, so you open the window, and now you get shit in your car as well as outside. I can show people who have shit on their car right now, but because it is not a 'significant' event, nothing happens. But when a staff person has it happen to them, something gets done. It most likely comes from the testing of the raw gas bleeders, so let's stop doing that.

. . .

Craig: That is always your answer, too difficult. I don't want anything done because nothing will be done, so forget it. You ask us to come here and care for the machinery, care for the company, sorry our company, but the company doesn't care for us. I am expected to go to the batteries and sell this, and they throw things like this back at us, and I don't have any answers. And one of our strategic objectives is to care for the employees, and they don't, so don't expect me to sell it.

The issue is a local one of tar deposits but in order to solve it there is a complex amalgam of interdependent technical and social elements. It raises questions of where the tar is coming from, and responsibility for 'leaking ovens' and 'operators leaving the doors open'. It then merged with broader issues of company lawyers recommending that management do not admit responsibility for such developments or it opened the plant up to a variety of legal suits. In addition, the discussion on recompensing plant employees who parked in the local plant car-park began to become intertwined with a broader issue of having to compensate people in other nearby car-parks who could make similar claims, and this was beyond the jurisdiction of local management. The issue of inequality also became dominant, as there were cases of superintendents who knew how to make claims for tar deposited from 'non-routine' emissions getting their cars detailed by the company, whereas operators who did not know about such regulations had their claims refused. In attempting to deal with the issues, it soon became clear that there was no common view; some operators perceived this as a crucial problem, while others did not regard it as a serious issue as they had come to terms with the fact that they worked in an environmentally unfriendly facility, a number of whom cycled to work or got a lift from a family member or colleague. The issue became used by some of the superintendents to illustrate the need for operators to be more responsible in running equipment and preventing environmental pollution, whereas a number of the operators saw it as an issue of management not investing adequately in new technology or taking responsibility when pollution did occur. This issue, as the quotation below reveals, has still not been resolved.

Craig: It's your tar.
Tom: It's our tar.
Craig: No it's not ours, I don't believe that any more.

Conclusion

As we have seen from the above discussion, implementation is an active, complex and emergent process in which both the technique or method being introduced and the production environment into which it is being introduced are part of a configurational process. The implementer has to address a whole range of complex issues as broad general techniques have their details and meaning interpreted in the local context, and action has to be taken to transform the local configurational process to incorporate the adapted change. In order to help address such issues, those working to implement change are involved in a collaborative and conflictual process of creating the necessary collective will for change and aligning the activities and interests of technology, operators and intrapreneurs in a new configurational process. In so doing, implementers inevitably act as collective designers and political entrepreneurs (Badham and Ehn 2000) in an ongoing process of reinvention and mutual adaptation. The difficulties faced, and the efforts and skills required, are those acknowledged by Pfeffer (1995), when he commented on the key importance for innovation of improving our ability to 'get things done' as compared to the traditional strategic issue of 'figuring out what to do'.

Note

1 Fleck emphasises the following: learning by doing (industrial learning curve), learning by undoing (reverse engineering), learning by using (incremental improvements in the course of using technology) cf. learning by trying (improvements necessary before a new configuration can be implemented) and learning by buying (acquire knowledge by hiring in expertise) and learning by learning (form expertise through in-house training programmes).

References

Abrahamson, E. (1991) 'Managerial fads and fashions: the diffusion and rejection of innovations', *Academy of Management Review* 16, 3: 586–612.

Bacharach, S.B., Bamberger, P. and Sonnenstuhl, W.J. (1996) 'The organizational transformation process: the micropolitics of dissonance reduction and the alignment of logics of action', *Australian Science Quarterly* 51: 477–506.

Badham, R. (1991) *Computers Design and Manufacturing: The Challenge*, Canberra: Department of Industrial Relations, AGPS.

Badham, R. (1992) 'Skill based automation: current European approaches and their international relevance', *Prometheus* 10, 2: 239–260.

Badham, R. (ed.) (1993a) 'Systems, networks and configurations', Special Issue, *International Journal of Human Factors in Manufacturing* 3, 1.

Badham, R. (1993b) 'Smart manufacturing techniques: beyond electronic drawing boards and turnkey systems', *International Journal of Human Factors in Manufacturing* 3, 2: 117–133.

Badham, R. (1995) 'Managing sociotechnical change: a configuration approach to technology implementation', in J. Benders *et al.* (eds) *Managing Technological Innovation*, London: Avebury Press, pp. 113–133.

Badham, R. (2000) 'Sociotechnical design', in W. Karwowski *et al.* (ed.) *Handbook of Human Factors and Ergonomics*, New York: Wiley, pp. 1031–1040.

Badham, R., Couchman, P. and McLoughlin, I. (1997) 'Implementing "vulnerable" socio-technical change projects', in I. McLoughlin and M. Harris (eds) *New Perspectives on Technology, Organization and Innovation*, London: Routledge, pp. 35–50.

Badham, R. and Ehn, P. (2000) 'Tinkering with technology: human factors, work redesign, and professionals in workplace innovation', *Human Factors and Ergonomics in Manufacturing*, 10, 1: 61–83.

Badham, R., McLoughlin, I.P. and Buchanan, D. (1998) 'Human resource management and cellular manufacturing', in N. Suresh *et al.* (eds) *Handbook on Cellular Manufacturing*, New Jersey: Prentice Hall.

Badham, R. and Wilson, S. (1993) 'Smart manufacturing techniques: beyond electronic drawing boards and turnkey systems', *International Journal of Human Factors in Manufacturing* 3, 3: 1–25.

Benders, J. and Van Veen, K. (2001) 'What's in a fashion? Interpretative viability and management fashions', *Organization* 8, 1: 33–53.

Braun, E. (1985) Constellation for manufacturing innovation', in E. Rhodes and D. Wield (eds) *Implementing New Technologies: Choice, Decision and Change in Manufacturing*, Oxford: Blackwell.

Buchanan, D. and Badham, R. (1999) *Power, Politics and Organizational Change: Winning The Turf Game*, London: Sage.

Buchanan, D. and Storey, J. (1997) 'Role taking and role switching in organizational change: the four pluralities', in I. McLoughlin and M. Harris (eds) *Innovation, Organizational Change and Technology*, London: International Thomson, pp. 127–145.

Charters, W.W. Jr. and Pellegrin, R.S. (1972) 'Barriers to the innovation process: four case studies of differentiated staffing', *Educational Agricultural Quarterly* 9: 3–4.

Clegg, S. and Hardy, C. (eds) (1999) *Handbook of Organizational Behaviour*, New York: Sage.

Daily Telegraph (2002) 'Leighton "hacked off" by Post Office politics', 19: 36.

Dunphy, D. and Stace, D. (1990) *Under New Management: Australian Organizations in Transition*, Sydney: McGraw-Hill.

Einjatten, F. van (1993) *The Paradigm that Changed the Workplace*, Stockholm: Arbetslivcentrum.

Elam, M. (1993) *Innovation as the Craft of Combination*, Linkoping: Department of Technology and Social Change, Linkoping University.

Fleck, J. (1999) 'Learning by trying: the implementation of configurational technology', in D. MacKenzie and J. Wajcman (eds) *The Social Shaping of Technology*, London: Open University Press, pp. 140–153.

Jervis, R. (1997) *System Effects: Complexity in Political and Social Life*, Princeton, NJ: Princeton University Press.

Law, J. (1997) 'TR2 passenger train', in J. Law *et al.* (eds) *Shaping Technology, Building Society*, Boston, MA: MIT Press, pp. 1–14.

Leonard-Barton, D. (1995) *Wellsprings of Knowledge: Building and Sustaining the Sources of Innovation*, Boston, MA: Harvard Business School Press.

MacKenzie, D. and Wajcman, J. (eds) (1999) *The Social Shaping Of Technology*, Milton Keynes: Open University Press.

McLoughlin, I., Badham, R. and Couchman, P. (2000) 'Rethinking politics and process in technological change', *Technology Analysis and Strategic Management* 12, 1, 1: 17–39.

Okamura, K., Orlikowski, W.J., Fujimoto, M. and Yates, J. (1998) 'Helping CSCW applications succeed: the role of mediators in the context of use', Unpublished Working Paper, MIT (Internet source: W.Orlikowski at wanda@mit.edu).

Orlikowski, W.J. and Hofman, D.J. (1997) 'An improvisational model for change management: the case of Groupware Technologies', *Sloan Management Review*, Winter, 11–21.

Perrow, C. (1983) 'The organizational context of human factors engineering', *Administrative Science Quarterly* 28: 521–541.

Pfeffer, J. (1995) *Managing with Power*, Boston, MA: Harvard Business School Press.

Rogers, E.M. (1995) *Diffusion of Innovations*, New York: Free Press.

Suchman, L. (1999) 'Working relations of technology production and use', in D. MacKenzie and J. Wajcman (eds) *The Social Shaping of Technology*, London: Open University Press, pp. 110–122.

Trigg, R.H. and Bodker, S. (1994) 'From implementation to design: tailoring and the emergence of systematisation in CSCW', *Proceedings of the ACM* 13, 2: 45–54.

Voss, C.A. (1988) 'Implementation: a key issue in manufacturing technology: the need for a field study', *Research Policy* 17, 2: 55–63.

6 Normalization of risks

A stream of bow visor incidents and the *Estonia* ferry accident

Hannu Hänninen

Introduction

Technological processes can have devastating outcomes. This chapter deals with a naval case where technological failures led to a serious crisis. It explores a stream of bow visor incidents in the Baltic ferry traffic and the related normalization of risks that led to the *Estonia* ferry accident. The Estonian-flagged ro-ro ferry *Estonia* sank in the northern Baltic Sea during the early hours of 28 September 1994. Out of the 989 people on board only 137 survived. The apparent reason for the capsizing was a bow visor failure. The *Estonia* ferry disaster may be the most tragic marine accident of our time. There had previously been a general belief that passenger ferries in the Baltic traffic were very safe, which made the disaster even more shocking. This kind of accident was not supposed to happen.

Interestingly, prior to the *Estonia* accident there had been several bow visor incidents on the Baltic ferries. Hence, what happened in the *Estonia* accident should not have been a total surprise. Different organizations and groups in the marine system that run and regulate the Baltic ferry traffic normalized the visor risk. The stream of visor incidents did not lead to significant improvements until it was too late. The bow visor attachments in the *Estonia* were not designed according to realistic assumptions; the impact of wave loads had been underestimated. Several questions arise: How did the unreliable and dangerous bow visor arrangement ever become a standard? Why did the marine system not learn from prior visor incidents? Why did no one question the dangerous ro-ro concept? How were some risks chosen and others ignored? This chapter aims to provide answers to these questions.

In previous research, there has been growing appreciation that accidents and disasters develop through minor failures and incidents (Turner 1978; Perrow 1984; Shrivastava 1987), and that warning signals may indicate possible crisis development. However, in previous explanations there has been little discussion of how organizational structure and culture guide risky decisions. Bureaucratic procedures, structural secrecy and culturally bound, rule-based behaviour direct the evaluation of risky technologies in organizations (Vaughan 1996, 1997). This chapter contradicts conventional failure explanations and argues that warning signals or deviance are usually clearly perceived and analysed. Therefore, accidents are often the result of conscious risk-taking. The bow visor incidents in the Baltic ferry traffic revealed the structural deficiencies of the visor concept. The aim here is to understand how the marine system and its organizational groups normalized the bow visor risk. The *Estonia* case also raises many questions concerning passenger safety. The study draws from the crisis, sociology of risk and technology literatures, which have usually been developed separately.

The *Estonia* and an error-inducing marine system

The Estonian-flagged *Estonia* was registered as a passenger/cargo ro-ro ferry (see Table 6.1) and was used for regular ferry traffic between Tallinn and Stockholm. Her overall length was 155.40 metres and she was able to carry 2000 passengers. The *Estonia* was built with a continuous vehicle-carrying space on the main deck. The entrance to the car deck was arranged through the bow. The installation comprised an upward-opening bow visor and a loading ramp that were common on the Baltic ferries at the time of her construction. Ro-ro ferries are easy to load and unload, but they are also very vulnerable. If for some reason water enters the car deck, the ferry becomes unstable and easily develops a list.

The *Estonia* capsized when large amounts of water entered the car deck after the bow visor locking devices had failed due to heavy wave loads. The bow visor, its locking devices and hinges failed under one or two wave impact loads. The visor moved forward, forcing the ramp partly open due to the mechanical interference inherent in the visor and ramp design. Water entered the car deck at the ramp sides. This led to a loss of stability and flooding of the accommodation decks. The full-width car deck made the list worsen rapidly. As a result of the increasing list and massive flooding, the vessel sank in less than an hour after the visor attachments gave in.

The Commission (Joint Accident Investigation Commission 1997), which investigated the accident, found that the bow visor attachments on the *Estonia* were not designed according to realistic assumptions; the impact of wave loads had been underestimated. Moreover, the attachments had been constructed with less strength than the calculations required. For a reason-

Table 6.1 The *Estonia*

The Estonian-flagged passenger/cargo ro-ro ferry.
Used for regular ferry traffic between Tallinn and Estonia.
Length 155.40 metres.
Able to carry 2000 passengers.
A continuous car deck with a entrance through the bow.
The bow arrangement includes a visor and a loading ramp.

Source: Joint Accident Investigation Commission (1997).

able level of safety the locking devices should have been several times stronger. The Commission also found that at the time of the *Estonia*'s construction, the industry's general experience of hydrodynamic loads on large ship structures was limited and design procedures for bow visors were not well established. It was also found that the information on prior bow visor incidents was not systematically collected, analysed and spread within the shipping industry.

Our examination about how bow visor risk was normalized illustrates how the whole marine system was behind the development of the *Estonia* accident. In fact, marine accidents are not rare, and the origin of problems seems to lie in the nature of the entire marine system. Perrow (1984) speaks of an 'error-inducing system' where 'the configuration of its many components induces errors and defeats attempts at error reduction' and that 'only a wholesome reconfiguration could make parts fit together in an error-neutral or error-avoiding matter' (Perrow 1984: 172). Accordingly, the system would not become much safer by improving or replacing any single component.

The main marine organizations (Table 6.2) include shipbuilders, classification societies, national authorities, international safety associations and shipping companies. These organizations have different roles in the marine system. They design and build ships, transport people and materials, insure the vessels and regulate the system.

Although marine transport has a long history, regulation is a relatively new part of the marine system. The earliest knowledge of ships comes from 6000 BC, so the system operated for almost 8000 years without regulation. Freedom of the seas has not formed a good foundation for international regulation. Since the late nineteenth century, maritime states have worked together for international regulations governing ship operation, design and safety. Accidents have fostered regulation. Yet there is not much federal presence in shipbuilding and the only international safety association is rigid and advisory. Many nations have only a little experience of national regulatory systems.

Acceptable risk

The *Estonia*'s visor failure may be linked to a stream of earlier visor failures. In order to understand the normalization of the bow visor risk, we need to

Table 6.2 Organizations in the marine system

Shipbuilders design and construct ships.
Classification societies set requirements and regulations and inspect vessels so that they can be insured.
National authorities guard local marine traffic and set local regulations.
The International Safety Association's (IMO) purpose is the improvement of maritime safety and it promotes the adoption conventions and protocols.
Shipping companies transport people, vehicles and goods. They normally own the vessels.

look at the marine system where the Baltic ro-ro ferries were designed, built and used, and where the critical decisions concerning the visors were made.

People make marine technology work. We may treat ship development, production and usage as a nondetermined and multidirectional process or flux whose direction is constantly negotiated and renegotiated among and between groups in the marine system (Bijker *et al.* 1987). Risk and safety are two of the key subjects in these negotiations. There is no such thing as absolute safety (Hollander 1997). Consequently, accidents are typically unfortunate outcomes of compromise. 'All technologies are shaped and mirror the complex trade-offs that make up our societies; technologies that work well are no different in this respect from those that fail. . . . Technologies always embody compromise' (Bijker and Law 1997: 3–4). Viewed in this light, the *Estonia* accident is an outcome of compromise and satisficing. With hindsight, it is obvious that the *Estonia* should have been designed differently to make it safer.

Politics, economics, theories of the strength of materials, professional preferences, skills and design tools among others all matter when a ship is designed and built. Technologies are shaped by a wide range of factors. Therefore, they might have been otherwise (Bijker and Law 1997: 3). Despite all uncertainties, technologies work more often than fail. Systems that carry risk potential pay special attention to reducing failures (Weick and Roberts 1993). This is why highly complex and tightly coupled systems such as nuclear power plants and air traffic controls are capable of running as safely as they do (Perrow 1984). The fact that the Baltic ro-ro ferries could have been 'otherwise' leads us to question why the ferries took the form they did, and why the visor concept was not changed even though it did not work properly. Answers will be found in the social, economic and technical relations of the marine system.

The development and use of marine technology as well as any technology involves constant negotiation and renegotiation among and between groups (Pinch and Bijker 1987; McLoughlin 1999). Once a temporary 'truth' has been winnowed from various interpretations, a 'closure' occurs. A closure requires emerging consensus among key social groups that a problem has been solved. In addition, risks are negotiated during technology development. Decision-makers form estimates of the risk involved in a decision. The estimates affect the risk actually taken. Even the most safety-conscious organizations have to rely on estimates. When organizations achieve a satisfactory level of safety, they stop searching for better alternatives (March and Simon 1958; Vaughan 1996). Ship development also includes such satisficing.

Many problems have to be solved before closure and stabilization. Various groups see arising problems differently. They decide differently about the definition of a problem and about the achievement of closure and stabilization (Pinch and Bijker 1987). Some of the problems that occur during technology development may persist and demand further development during use (Hughes 1987). There are also often problems that technology developers could not foresee. Technology developers have to accept risks during the development stage and some of these risks may be realized afterwards in adverse circumstances or because they were assessed in the wrong way at the beginning. The unsolved problems are warnings showing that there is something wrong with the system. In the *Estonia* case, the prior visor incidents signalled that the visor concept did not work well and that the force of the wave loads had been underestimated. Baltic ferry constructors had not anticipated that visor problems would emerge. When these problems began piling up they nevertheless continued to construct a similar type of visor for several years until the concept was gradually altered.

Although previous crisis studies (e.g. Turner 1978; Fink 1986) have described warning signals and crisis symptoms as deviations from 'normal' activity, I take a different stance and argue that warnings occur fairly often. Technological systems are not error-free. Minor disturbances and failures are common and they are normally fixed before anything serious happens. Despite the common conception, people in organizations often perceive warning signals well. They interpret warnings and respond to risk when they react to the deviance. Hence crises may develop through normal decisions.

Organizational context guides how warnings are dealt with. Routines and systems can overwhelm well-intentioned individuals, and normal everyday practices, if carried out without further thought, may produce failures. In risk perception, people behave less as individuals and more as social beings who have internalized social pressures and delegated their decision-making processes to institutions. 'They manage as well as they do, without knowing the risks they face, by following social rules on what to ignore: institutions are their problem-simplifying devices' (Douglas and Wildavsky 1983: 80). A decision-making process provides an occasion for executing standard operating procedures, and fulfilling role expectations and duties (March and Olsen 1976; March 1999).

Risky decisions can also be looked at as rule-based actions. Viewed in this light, decision logic stems from the logic of appropriateness, obligation, duty and rules (March 1999). People seek to fulfil their identities as organizational actors in decision-making. Rules evolve over time, and current rules store information generated by previous experience. How the rules are formed depends greatly on the way previous experience is interpreted either as success or failure. Further, rules are clarified by reference to comparable rules; situations are clarified by reference to analogous situations (March 1999: 25).

Systems that carry accidental potential construct experience bases (Huber 1991) upon which new safety-related decisions are reflected. Positive experiences strengthen the bases so that any deviance from acceptable standards is viewed as a signal of potential danger. Past problem definitions and past methods of responding to the problems become part of the social context of decision-making (Vaughan 1996). Positive experiences increase the propensity to risk-taking. Warnings can be normalized. Understanding risky decisions requires knowing the organizational contexts of those decisions. Despite the cumulated experience and stored knowledge of risks, relative safety is rather a dynamic product of learning from error over time (Douglas and Wildavsky 1983).

Risk negotiations cannot be avoided during ship development and usage. Different groups agree on risks, define acceptable risks and choose some risks over others. The term 'acceptable risk' (Clarke 1989; Heimann 1997) refers to risk negotiation. Accepted risk is also an inherently political issue. The word 'acceptable' implies a political, not a scientific judgement (Clarke 1989). Understanding safety-related decisions in marine transport requires understanding the organizational and political contexts of those decisions.

What the experts call 'actual' risk estimates are based on the presupposition that the probability and consequences by which the risk is measured can be quantified. Risk assessment may be considered a technical procedure, which is to be undertaken through rational calculation of ends and means (Fox 1999). Within such a stance (e.g. Johnstone-Bryden 1995; Heimann 1997), 'all risks may be evaluated and suitably managed, such that all may be predicted and countered, so risks, accidents and insecurities are minimized or prevented altogether' (Fox 1999: 13). However, even at their best, risk estimates are very rough and imprecise. 'Knowledge of danger is necessarily partial and limited: judgements of risk and safety must be selected as much on the basis of what is valued as on the basis of what is known' (Douglas and Wildavsky 1983: 80–81). Objective evidence about technology is not enough; acceptable risk is a matter of judgement (Douglas and Wildavsky 1983).

Although, in the light of previous technology literature, the bow visor technology in the Baltic ferries is not very complex, it is still subject to risk and uncertainties. For vessel designers the amount of risk could not be known for sure. No matter how tightly the visor seals the bow it will still leak a little. The construction would have to be extremely strong if it was not to any extent vulnerable to wave loads. In fact, the O-ring sealing problem that caused the Challenger space shuttle disaster is somewhat similar to the visor problems of the Baltic ro-ro ferries. NASA acknowledged that the joint did not work as per the design, but they felt that it worked well enough to be considered an acceptable risk (Vaughan 1996). The bow visors in the Baltic ro-ro ferries do indeed leak small amounts of water on to the cardecks, but not so much that it would be a safety concern. The analogy to the space shuttle case comes from the fact that a bow visor is never as safe as a solid bow. What kinds of wave load should the visor and its attachments sustain to be considered safe? How safe is safe enough?

If a decision is made to use a risky concept, then the acceptable level of risk has to be defined. The problem is often that technology experts and scientists tend to put risky decisions and solutions into technical frames (Clarke 1989). For technical experts, risk assessment involves highly technical issues, data and tests. Yet, by answering technical questions in any safety-related problem, technical experts answer or rather decide social problems at the same time. Social problems are often defined and solved as 'technical' problems.

Clarke (1989) examined the 1981 contamination of an office building in Binghamton. In this compelling case study he generated insights into the paradox of acceptable risk. The case illustrates how the 'real problem' was neither the contamination itself nor cleaning the building, but convincing people that the building is safe to re-enter. Clarke's (1989) study describes how organizations constantly balance between risk and safety, and how they try to convince themselves and others that the risks taken are acceptable (see also Clarke 2000). They are often unwilling to share their knowledge of the risk with the public. The claims of 'almost absolute safety' are common. MacKenzie (1990) has uncovered some of the politics that make up the social construction of complex technology. He has described processes where ballistic missiles are made to look more accurate and hence nuclear-weapons technology much safer than it actually is (see also Sagan 1993). As different assessments, tests and calculations of technology and its safety become public, they also become more 'objective' and more 'real' (Berger and Luckmann 1966). The shipping companies who run the Baltic ferry traffic had, through advertising, created an image of absolutely safe luxury cruises. The public, however, was unaware of the visor risk.

The normalization of risks

Organizational groups perceive risks well. Usually, the surprise element in crises is not so much related to unexpected failures, but rather to whether risk assessments will hold. Therefore, the question is not whether people know about technology problems and risk. Generally, people know what they are supposed to know. We need to ask instead what is their construction of risk and how they construct risks (Vaughan 1996).

The problem with many crisis analyses is that they ignore history. This is because they focus only on the critical decisions that led to the crisis, without taking into account the stream of cases or the overall caseload that has affected these decisions. People tend to evaluate a problem in relation to some larger problem set, usually the sequence of previous cases. Decision-making is serial and problems are solved in sequence. Previous cases are even more useful if they include other decisions on the same problem. 'A sequence of decisions made about a given case, or issue creates a decision stream' (Vaughan 1996: 243). 'The selection of risks worth taking and avoiding is made by a process, not by a person' (Douglas and Wildavsky 1983: 93).

If a warning or incident occurs in a technological system, it is treated as information that deviates from expectations. Each incident is evaluated within the decision stream of related cases. As the experience of earlier cases accumulates incrementally, the decision-makers may end up with conclusions that are difficult for outsiders to understand (Vaughan 1996). It is easy to judge decision-makers without getting to know the context in which the decisions were made.

Following Vaughan (1996), the sequence of similar events may originate a process where risks are eventually normalized. Frequent events lose some of their seriousness when they occur in sequence. Methods of assessing and responding to them tend to stabilize and routines easily develop. 'Knowledge and expectations about performance that are born out time after time encourages decision-makers to classify each similar event as a normal or "typical" case of X' (Vaughan 1996: 246). Viewed in this light, even serious warnings may be interpreted as routine signals, the effects of which can be predicted. This does not mean that decision-makers would ignore warnings. They merely attach them to the stream of earlier warnings.

The interpretation of the first incident plays an important part in a stream of incidents. The first incident may be defined as a serious warning or just a normal deviance. Once an anomaly has been accepted it is also likely to be accepted the next time. After a few times, a routine has developed. Thereafter, it is difficult to go back and reassess the risk. The groups that have accepted an anomaly often feel pressure to continue risk-taking. It is also tempting to accept more risk if solving a problem is difficult and if previous risks have not led to failure.

Thus I have developed the argument that in many cases, decision-makers are familiar with the problems that eventually cause a crisis. The processes in crisis development may be traced back to the drawing tables of technology inventors and developers. Some problems often persist after the closure and new ones emerge when use begins. Technology developers and users have to accept risks that may be realized as failures in adverse circumstances. Within this kind of framework we can distinguish different phases from crisis development. The development may follow the pattern that is presented below. The emphasis here is on decision-making routines and rule-following which may initiate vicious circles (Masuch 1985) where well-intended acts can produce great harm.

1 *Technology is invented and developed.* Technology invention and development are wasteful processes. Several variants are created; some of them die and others survive. Social groups in the development process decide which problems are relevant. They recognize some risks and ignore others.
2 *Closure and stabilization.* Closure in technology development involves stabilization and the disappearance of problems. Safety problems are often not fully solved. Instead, it is enough if the relevant social groups consider them to be solved. Claims of 'almost absolute safety' are common, although the safety problems may be well known to engineers (Pinch and Bijker 1987).
3 *Unexpected problems and deviance.* Despite careful planning, all problems cannot be foreseen. Technologies are refined and improved during use.

People develop useful ways to handle technology and its problems. Work groups develop local routines and rules as experiences accumulate. They develop specific ways of proceeding and shared definitions of repeated situations. Work group cultures are developed out of these collectively constructed realities (Vaughan 1996).

4 *Perceived escalated risk.* Organizations perceive operational failures and take appropriate action. If the unexpected problems or deviance are serious, the organization has to react and take official action where the escalated risk is acknowledged (Vaughan 1996).
5 *Analysis of evidence.* Decision-making easily becomes patterned. The patterns that have shaped decision-making in the past are reproduced in new situations (Vaughan 1996). People try to avoid mistakes but may be blind to the structural and cultural binds that guide their behaviour. Moreover, they struggle with limitations in attention, memory and comprehension. Decision-makers are susceptible to satisficing; they see what they expect to see and overlook unexpected events (March 1994). They analyse new evidence in the light of past performance but retain more optimism than is justified by that experience. Hence, success may contain the seeds of failure (March 1994).
6 *Accepting risk.* Decision-makers may evaluate past performance and success with optimism. They may accept more risk and deviance than was originally planned. This is more likely to happen if the problems are complicated or if the costs are too high. If the safety improvements require changing a widely used standard, they are not likely to happen.
7 *Risky attempts.* After accepting risk and deviance, an organization may continue its activities with minor repairs or modifications. The risk-taking is now conscious. Often only some work groups are aware that the whole system is operating in the area of accepted risk. If the system fails, it surprises many.
8 *Failure.* Risk-taking does not necessarily lead to failure but when it does, a crisis may emerge. Minor incidents or deviation add to the experience base and the circle starts all over again. Only significant losses lead to thorough evaluation of the whole system.

Although the sequential process here was framed with the *Estonia* accident in mind, this kind of pattern or parts of it may be found in the background of other accidents as well. Next we shall go through these phases with empirical details on the *Estonia* accident. First, we look into the ro-ro and visor concepts and their history and explain why they became so widely used. We then go into the history of the Baltic ro-ro ferry traffic. It began between Finland and Sweden where the ro-ro and visor concepts developed into a standard and continued later between *Estonia* and Sweden. It is necessary to understand this development and the overall circumstances of the *Estonia* accident before we elaborate on the stream of earlier visor incidents and the normalization of risk.

The stabilization of ro-ro and bow visor concepts

Ro-ro vessels are constructed for carrying cargo which can roll on and off. They have open internal decks and they are usually equipped with stern and bow ramps to accommodate loading and unloading. Although the overall concept is very simple it offers a flexible approach to cargo handling. Despite its obvious benefits the ro-ro technique also requires a greater degree of seamanship and expertise in operation than do conventional techniques. When the ro-ro concept started to spread, the ro-ro and visor technology soon became stabilized also in Baltic marine transport.

The overall architecture of ro-ro vessels has not changed since the 1960s. Within ro-ro technology there were alternative bow arrangements where the closure was not so obvious. The access to internal deck(s) via a forward ramp has been solved mainly in two ways in the ro-ro concept. The outer enclosure of the bow opening is arranged either as a pair of clam doors or as a visor, as on the *Estonia* (see Figure 6.1), hinged at the upper deck and opening upwards. Under rough sea conditions the visors may be exposed to sea loads in the opening direction.

The ro-ro ferry traffic between southwest Finland and the Stockholm region in Sweden began around 1960. The growth rate of this traffic was extraordinarily fast. New ships were built at such a rapid rate that regulation could not keep pace with development. The size, capacity and comfort of the ferries as well as the number of passengers and vehicles carried

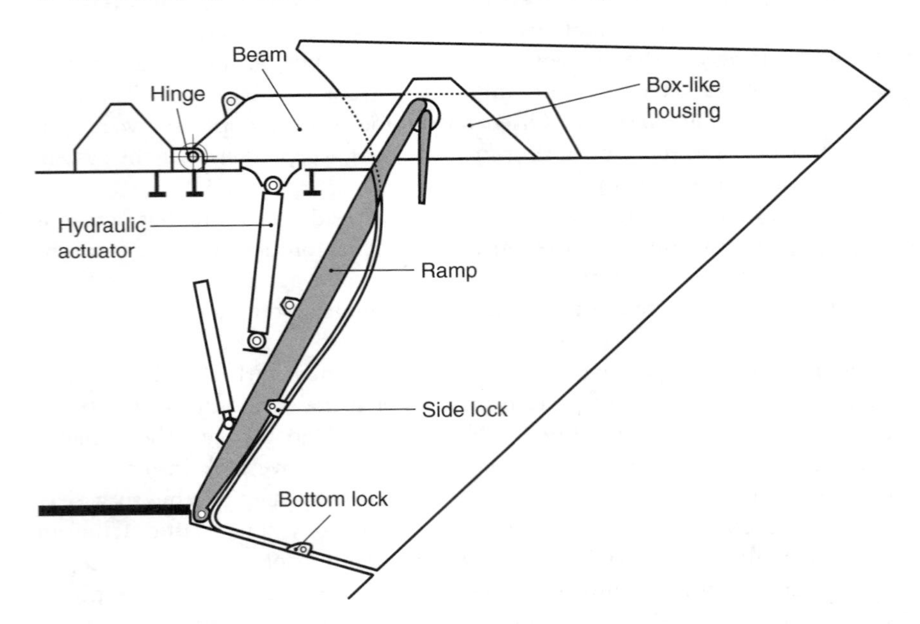

Figure 6.1 The *Estonia*'s bow visor and ramp installation

Source: Joint Accident Investigation Commission (1997)

increased rapidly. The development was geared up by the competition between two major shipping company groups, Viking Line and Silja Line. The companies soon found the ro-ro concept very useful and as a result purpose-built ro-ro ferries were ordered. The ro-ro concept proved successful and was quickly established as a standard and an indispensable transport element. The success of ferry traffic between Finland and Sweden made it lucrative to apply the idea later to the passage between Estonia and Sweden. Traffic between Estonia and Sweden was first carried by the ro-ro ferry *Nord Estonia* until she was replaced by the *Estonia* in February 1993. The *Estonia* was built originally for ferry traffic between Finland and Sweden.

The shipping companies considered the ro-ro concept a success and did not hesitate to use it for all Baltic ferry traffic. As no serious accidents occurred, closure and trust in the ro-ro concept were confirmed. Large size and efficiency were considered important. The technical development of ro-ro ferries proceeded on the terms of shipbuilders. The pace of this development was so fast that the regulators were not able to keep up with it. There were no remarkable variations to ro-ro technique and how it was applied, other than that the vessels had two kinds of bow arrangement: clam doors and visors. There were three variations of the visor concept: (1) an independent visor with no housing for accommodating the top of the ramp, (2) a visor and box-like housing for the stowing ramp as in the *Estonia*, or (3) a visor and box-like housing for the stowing ramp and a separate barrier. Bow visors were not built after 1988. The Baltic ferry traffic was ready to experiment with the ro-ro technique from the beginning of its introduction to commercial marine transport. The Baltic marine system developed a local standard from the basic ro-ro concept. Different bow arrangements became alternatives within this standard.

The stream of visor incidents

As we already know, the application of ro-ro technology caused problems. Groups in the marine system had put the bow visor incidents aside because they did not believe in the possibility of a major accident. In retrospect, the incidents seem very serious.

According to previous literature, there are usually more or less visible warning signals before the actual crisis is triggered. Previous crisis studies have shown that a typical crisis is preceded by two or three similar types of events or a few different types of events that can be identified as early warnings. What makes the *Estonia* case special is the fact that there were several similar incidents that preceded the final failure. These incidents, some of which may be considered to be accidents, are listed in Table 6.3.

An incident occurred also to the *Diana II* on 14 and 15 January 1993. The *Diana II* was a near-sister vessel to the *Estonia* and hence its visor design was similar to that of the *Estonia*. The visor locking arrangements of the *Diana II* were damaged when the starboard locking device lug was lost, the

Table 6.3 The stream of visor incidents in the Baltic ferry traffic

Visby, December 1973
Stena Sailer, January 1974
Svea Star, May 1974
Wellamo, December 1975
Saga Star, May 1982
Viking Saga, October 1984
Stena Jutlandica, October 1984
Ilyich, December 1984
Mariella, November 1985
Tor Hollandia, 1985–1986
Finnhansa, January 1977
Diana II, January 1993
Silja Europa, September 1994

Source: Joint Accident Investigation Commission (1997).

bottom log bent and its weld cracked. The damage was repaired by normal procedures to conform to the original standard. The classification society in charge, Bureau Veritas regional office in Gothenburg, did not consider the incident serious enough to investigate the matter further. Neither was any initiative taken for further actions (Joint Accident Investigation Commission 1997).

The visor incidents should have indicated that the visor concept, primarily the visor attachments, had serious deficiencies and did not work as per the design. None the less, groups in the marine system failed to interpret these warning signals as a substantial threat. Information on the visor incidents was not spread within the industry and these incidents were never discussed widely within the marine system. The reporting system did not work. The groups in the marine system had to react to the incidents whose consequences had to be evaluated and repaired. The marine system had to face the visor risk because the visor damages were undeniable and at least required to be repaired to meet the original standard before the ferries could be put back into service.

The normalization of visor risk

Although the information about the earlier visor incidents was neither systematically collected nor spread within the marine system, the authorities filed any evidence of incidents that they happened to receive. There was not much reporting required in the system, and even the few reporting rules were often neglected. Shipping companies, yards, classification societies and local authorities all took turns in handling the visor incidents and the outcome was always the same: to continue without any major changes.

Normally, information on a new incident reaches the shipping company first. Mariners and shipping companies are responsible for reporting all incidents to local authorities. Shipping companies also report to classification societies about significant incidents that occur to vessels classified by them. Classification societies are bound to professional secrecy, and hence cannot pass on information directly from shipping companies. Although classification societies analyse classified vessels and develop rules, they need to preserve the anonymity of shipping companies. Shipyards receive the information on a possible incident only if it happens during the one-year warranty when they are responsible for the repairs. Subcontractors, who design and manufacture bow visors, are alerted by the shipping companies to check the situation after visor failures or incidents.

The reporting of the earlier incidents did not fulfil all requirements. Local authorities, in particular, were not always notified. For example, they were never advised of the *Finlandia* and the *Mariella* incidents. On the other hand, there are examples of good information exchange as well. The evidence of the *Wellamo* hazard, for instance, reached the yard, the authorities and the classification society. Sometimes the different parties discussed the incident together. The *Mariella* bow failure was discovered to be so interesting that Wartsilä Yard, Navire (the visor subcontractor), and Norske Veritas (the classification society) discussed the evidence together. In cases such as this, the shipping company is responsible for decision-making, and if the incident happens during the warranty period the yard is responsible as well.

As already mentioned, the reporting system did not work smoothly. However, there was no way to avoid facing the risk involved in these incidents. Every incident was discussed in the work groups that had the evidence and on some occasions it was decided that the evidence would be passed on, but often this was not the case. Immediately after the *Estonia* accident there were personnel at shipyards who knew that the cause had been a bow visor failure. They were well aware of earlier incidents. The work groups had acknowledged the visor problem and treated it as an acceptable risk.

There are no records of meetings held where the bow visor incidents and related risk were discussed. The incidents were handled as a part of regular decision-making. Fortunately, we have documents that indicate which of the involved organizations knew of each incident and the outcomes of their decisions. In part, the hazards did not attract more attention because it is common for vessels to suffer damage in rough weather; bulwarks and gunwales bend, machines on the deck break, windows break, and bulkheads bend inwards. The bow visor problem was just one of the many others, and the decision-makers failed to define them as a weak spot. In addition, the classification societies involved estimated the bow visor risk differently. In 1978, Germanischer Lloyd had a specific formula for the design load of a bow visor that gave three times the load used for the *Estonia*. A co-operative organization, the International Association of Classification Societies (IACS), was somewhat concerned about the bow visor risk because it defined the pressure head and the calculation procedure to be applied more clearly. It also issued more detailed rules in the 1982 Unified Requirements and in later recommendations. However, according to the 'Grand Father Rule' these requirements were not applied to the *Estonia* and other older vessels (Joint Accident Investigation Commission 1997).

In some cases, the hazards led to improvements. The visor attachments were strengthened after negotiations between the classification societies and shipping companies. Alas, the incidents were treated as normal problems and no one in the system came forward demanding greater attention to the issue. The ro-ro and bow visor concepts were not challenged. The near misses (March *et al.* 1991) were not enough to capture the decision-makers' attention. The history was considered successful. The past performance and 'success' indeed contained the seeds of failure.

The bow visor concept in the Baltic ro-ro ferries was an accepted risk. The effects of sea loads are difficult to estimate. Even so, the industry had experienced numerous visor failures prior to the *Estonia* accident, which was enough to show that the bow visor concept included serious unsolved problems.

The organizations in the marine system played different roles in the normalization of the visor risk. The *shipping companies* rarely had any significant improvements made in the visor system. At times, they did not report the incidents to the authorities. They used the ferries to transport people and vehicles, and had invested a lot in the ferry standard. In fact, the whole business idea was based on the standard. The *classification societies* occupied a dubious role in the process. They were safety authorities that checked the vessel so that it could be insured. However, they also tried to serve their clients: the shipping companies. At first, the classification societies did not have enough information and experience of the real wave loads. Hence, their visor requirements varied significantly. Later, they developed new requirements that were not applied to older vessels. In the end, they did not consider the risk serious enough to try to draw wider attention to the matter. The *yards* basically built what was ordered by the shipping companies, simultaneously trying to fulfil the requirements of the classifications societies. The ferry technology development processes involved negotiations between the yards and shipping companies and so the yards were not active in the normalization of risk, yet neither were they passive actors. They received information about the near misses at times. The *local authorities* were authorized to influence the visor attachment requirements of the vessels under their own flag and also of vessels elsewhere through the 'port state control' inspections. Such influence did not, however, take place. The data of the reported visor hazards were filed in the archives.

Many groups involved accepted the visor risk. In a way, the whole marine system shared the risk. Yet not all groups knew or were informed of the risk. Unfortunately, for example, the *Estonia* crew were not told about the visor risk. The information was not shared because the risk was not considered to be anything unusual.

The *Estonia* accident revisited

We now close the *Estonia* case by looking at how the accident developed (Table 6.4). The phases leading up to the accident are seen as parts of a coherent chronological story. We have placed the *Estonia* accident in a sequence of other bow visor failures in the Baltic ferry traffic. The faulty visor designs have been traced back to the development of ro-ro technique and the bow visor concept. We have already described how the ro-ro and bow visor concept were developed and how they were stabilized in the Baltic ferry traffic. It is clear that the real wave loads on the visors were not understood. Visor incidents began to pile up. Yet this did not lead to criticism of the visor concept. Later, when wave loads were understood better, new visor requirements took effect, but were not applied to old vessels. The marine system reacted to the visor incidents by accepting the visor risk. The bow visors in the Baltic ferries were defined as an acceptable risk. The visor risk was negotiated and normalized in each case by groups who had information on the failure. The matter was never discussed openly within the marine system.

Discussion and conclusions

The reported study warrants conclusions on three issues. First, with respect to the specific accident itself, we would like to emphasize a few points. With good luck, the *Estonia* accident might have been avoided, although it is

Table 6.4 The *Estonia* accident revisited

1	*Technology is invented and developed* The ro-ro technique is developed as an efficient and convenient form of marine transport.
2	*Closure and stabilization* The ro-ro and bow visor concepts are widely and rapidly accepted. Vessels are built for the Baltic ferry traffic. Wave loads are not understood.
3	*Unexpected problems and deviance* Visor incidents begin to occur.
4	*Perceived escalated risk* Each new incident has to be repaired.
5	*Analysis of evidence* Visor risks are negotiated in each incident separately. There are no public discussions about visor risk. Information of the incidents is not shared.
6	*Accepting risk* The stream of incidents leads to only minor improvements. New visor requirements are not applied to old ferries, however.
7	*Risky attempts* The visor concept is not questioned. Visor risk is normalized and ferry traffic continues.
8	*Failure* The *Estonia*'s visor fails under wave loads. Water enters the car deck and the vessel capsizes. The visor risk was realized.

evident that the marine system took excessive risks when it evaluated the earlier incidents and the workings of the visor concept. The incidents should have set bells ringing. Yet vessels often suffer damage in rough weather, and therefore the visor incidents were buried among other problems. In addition, the ro-ro and bow visor standards that formed the basis of the Baltic ferry traffic guided risky decisions and were rigid to change. After the incidents, and even after the *Estonia* accident, the Baltic ferry traffic still relies on the ro-ro standard. However, in the wake of the accident the system abandoned the visor concept and now uses mainly clam doors on ro-ro vessels.

It may often take a disaster to change the state of affairs. However, it is surprising how little the *Estonia* crisis affected Baltic ferry traffic. Since the accident a lot of work has been done to improve ferry traffic safety. The improvements include new bow visor requirements, visor inspections and new weather recommendations. A Baltic ro-ro ferry trip is now safer than ever before. However, the bow visor incidents have not disappeared. We have seen a few incidents since the accident and perhaps we shall see more.

Second, with respect to the marine system and passenger safety, the *Estonia* case and previous marine accident studies justify the claim that production pressures in the marine system tend to jeopardize passenger safety. The structure of the industry with its insurance and regulation difficulties makes the system insensitive even to clear warnings. The *Estonia* case also supports Perrow's (1984) argument that the marine system is indeed an error-inducing system. Unfortunately, technological developments and attempted fixes are not likely to succeed in this system in the near future. The structure of the industry and insurance as well as the difficulties of national and international regulation form the basis for resistance to new solutions. Only a profound reconfiguration of the system would make the parts fit together in an error-avoiding matter (Perrow 1984). One needs to overcome the structural resistance and tradition of error inducement in order to develop necessary regulation for marine traffic and transportation.

Third, with respect to lessons to be learned, the stream of visor incidents in Baltic ferry traffic exemplifies the process-like descriptions of crisis development. Yet the numerous errors and incidents in marine traffic indicate against previous descriptions of warnings as rather rare events. The stream of incidents in this study was special because they were so similar. Nevertheless, the fact that there were many warnings is not unusual. Technologies do not work without problems, and if they are not fixed, the preconditions for failure development are created. Another point that needs to be emphasized is that crises often develop through conscious risk-taking. Organizations perceive warnings well and make decisions about how to handle them. Risks are negotiated. Accidents occur not because the warning would have gone unnoticed, but because the risk assessments have not held.

A further lesson from the *Estonia* case is that the weight of a standard (Schmidt and Werle 1998) can never be overestimated. Much of the Baltic ro-ro ferry traffic was based on the same standards. Unexpected problems are a part of technology development. The systems are redefined and improved through use but if the improvements call for changing an essential standard, they are not likely to happen. Hence even clear warning signals do not necessarily lead to changes in the system. Investments in the Baltic ferry traffic were substantial and there were undeniable production pressures, which promoted the marine system's commitment to existing ro-ro and bow visor standards.

The study also shows that different phases in ship design, building and usage all call for different ways of dealing with risks. It seems that the marine system is slow to react or cannot handle risks during usage. It is impossible to predict all problems, but when they do occur the system should react to them. One central cause for the normalization of risks in the *Estonia* case was the frequent damage that vessels suffer in rough weather. If the amount of storm-caused damage could be reduced, the important warnings might have more impact.

Finally, this study suggests the inability of communities such as the marine system to learn from failures. This has a lot to do with lack of communication. Incident reporting and collective risk assessment would facilitate dialogue among marine organizations. Groupthink in decision-making may be avoided by making the groups more heterogeneous. Risk experts without technology expertise could be brought in to secure multiple perspectives in the collective interpretation of risks. Laypeople and the public often view risks differently from the technology experts who make the risky decisions. In the *Estonia* case, the stream of visor incidents could have led to a thorough risk evaluation sooner had the decision-makers represented different perspectives and professions.

References

Berger, P. and Luckmann, T. (1966) *The Social Construction of Reality*, New York: Doubleday.

Bijker, W.E., Hughes, T.P. and Pinch, T.J. (1987) 'General introduction', in W.E. Bijker, T.P. Hughes and T.J. Pinch (eds) *The Social Construction of Technological Systems: New Directions in the Sociology of History and Technology*, Cambridge, MA: MIT Press.

Bijker, W.E. and Law, J. (eds) (1997) *Shaping Technology/Building Society – Studies in Sociotechnical Change*, Cambridge, MA: MIT Press.

Clarke, L.B. (1989) *Acceptable Risk? Making Decisions in a Toxic Environment*, Berkeley: University of California Press.

Clarke, L.B. (2000) *Mission Improbable: Using Fantasy Documents to Tame Disasters*, Chicago, IL: University of Chicago Press.

Douglas, M. and Wildavsky, A. (1983) *Risk and Culture – An Essay on the Selection of Technological and Environmental Dangers*, Berkeley: University of California Press.

Fink, S.J. (1986) *Crisis Management: Planning for the Inevitable*, New York: American Management Association (AMACOM).

Fox, N.J. (1999) 'Postmodern reflections on "risk", "hazards" and "life choices"', in

D. Lupton (ed.) *Risk and Sociocultural Theory – New Dimensions and Perspectives*, Cambridge: Cambridge University Press.

Heimann, C.F.L. (1997) *Acceptable Risks – Politics, Policy, and Risky Technologies*, Ann Arbor, MI: The University of Michigan Press.

Hollander, R.D. (1997) 'The social construction of safety', in K. Shrader-Frechette and L. Westra (eds) *Technology and Values*, Lanham, ML: Rowman & Littlefield.

Huber, G.P. (1991) 'Organizational learning: the contributing processes and the literatures', *Organization Science* 2, 1: 88–115.

Hughes, T.P. (1987) 'The evolution of large scale technical systems', in W.E. Bijker, T.P. Hughes and T.J. Pinch, (eds) *The Social Construction of Technological Systems: New Directions in the Sociology of History and Technology*, Cambridge, MA: MIT Press.

Johnstone-Bryden, I.M. (1995) *Managing Risk*, Aldershot: Avebury.

Joint Accident Investigation Commission of Estonia, Finland and Sweden (1997) *Final Report on the Capsizing on 28 September 1994 in the Baltic Sea of the Ro-ro Passenger Vessel* MV ESTONIA, Helsinki: Edita.

MacKenzie, D. (1990) *Inventing Accuracy: A Historical Sociology of Nuclear Missile Guidance*, Cambridge, MA: MIT Press.

McLoughlin, I. (1999) *Creative Technological Change – The Shaping of Technology and Organisations*, New York: Routledge.

March, J.G. (1994) *A Primer on Decision Making: How Decisions Happen*, New York: The Free Press.

March, J.G. (1999) *The Pursuit of Organizational Intelligence*, Oxford: Blackwell.

March, J.G. and Olsen, J.P. (1976) *Ambiguity and Choice in Organizations*, Oslo: Universitetsforlaget.

March, J.G. and Simon, H.A. (1958) *Organizations*, New York: John Wiley.

March, J.G., Sproull, L.S. and Tamuz, M. (1991) 'Learning from samples of one or fewer', *Organization Science* 2, 1: 1–13.

Masuch, M. (1985) 'Vicious circles in organizations', *Administrative Science Quarterly* 30: 14–33.

Perrow, C. (1984) *Normal Accident: Living with High Risk Technologies*, New York: Basic Books.

Pinch, T.J. and Bijker, W.E. (1987) 'The social construction of facts and artifacts: or how the sociology of science and the technology might benefit each other', in W.E. Bijker, T.P. Hughes and T.J. Pinch (eds) *The Social Construction of Technological Systems: New Directions in the Sociology and History of Technology*, Cambridge, MA: MIT Press.

Sagan, S.D. (1993) *The Limits of Safety – Organizations, Accidents, and Nuclear Weapons*, Princeton, NJ: Princeton University Press.

Schmidt, S.K. and Werle, R. (1998) *Coordinating Technology – Studies in the International Standardization of Telecommunications*, Cambridge, MA: MIT Press.

Shrivastava, P. (1987) *Bhopal: Anatomy of a Crisis*, Cambridge, MA: Ballinger.

Turner, B.A. (1978) *Man-made Disasters*, London: Wykeham.

Vaughan, D. (1996) *The Challenger Launch Decision: Risky Technology, Culture, and Deviance at NASA*, Chicago, IL: University of Chicago Press.

Vaughan, D. (1997) 'The trickle-down effect: policy decisions, risky work, and the *Challenger* tragedy', *California Management Review* 39, 2: 80–102.

Weick, K.E. and Roberts, K.H. (1993) 'Collective mind in organizations: heedful interrelating on flight decks', *Administrative Science Quarterly* 38: 347–381.

7 Kicking against the pricks[1]

Corporate entrepreneurship in mature organizations

Oswald Jones

Introduction: managing maturity

The end of the millennium was marked by a widespread obsession with the dot.com phenomenon typified by companies with few tangible assets and little likelihood of profits. The bursting of the dot.com 'bubble' illustrates the importance of continuing commitment to companies which represent the 'old' economy. MFD, a mid-sized manufacturing company, which is the subject of this chapter, is important for a number of reasons. First, MFD is undergoing a transformation from a batch producer of engineering components to the mass manufacturer of electronic assemblies. Second, the company is based in a small, relatively isolated town in which it is the major employer (400 employees) and of massive significance to the local economy. Third, the case illustrates the interaction of corporate entrepreneurship and institutional change. MFD, a privately owned company, was founded over forty years ago to supply casting and machined components to the Ministry of Defence (MoD). An inability to obtain new markets to replace defence-related work meant that between 1992 and 1994 the workforce was reduced by 50 per cent to approximately 450 employees. The owner/manager who was increasingly concerned about the company's ability to survive initiated a number of changes including the recruitment of managers with experience of modern manufacturing and marketing techniques. Gradually, a range of new technologies were introduced which were used to develop new products, improve manufacturing processes and create better communication systems within MFD. At an operational level, the change agent was a middle manager who, in contrast to most other long-serving white and blue-collar workers, joined the company less than one year before the change programme began. He drew on his experience of mass-production techniques to instigate a range of changes which revolutionized manufacturing activities within the company.

Change management has attracted a considerable amount of academic interest since the emergence of a field of inquiry termed 'organization theory'. Beginning with the seminal work of Lewin (1947) and including important contributions by Bennis (1964) through to more recent work

(Eden and Huxham 1996; Hartley *et al.* 1997) the concept of a 'change agent' has been a central feature of academic research. Primarily because of the North American influence terms such as 'intrapreneur' (Pinchot 1985) or 'corporate entrepreneur' (Burgelman 1983) are now more common in the management literature. One consistent theme in such work is that corporate entrepreneurs are viewed as 'social deviants' prepared to break organizational rules (kicking against the pricks) in implementing change programmes (Shane 1994). A concurrent body of literature examines ways in which values, norms and beliefs influence both managerial decisions and organizational structures (Meyer and Rowan 1977; Powell and DiMaggio 1991). Recently, a number of authors have argued that institutionalists must provide a fuller account of individual agency in explaining the process of change (Fligstein 1997; Beckert 1999). The interplay of agency and structure as described in structuration theory (Giddens 1984) provides the metatheory for this account of change management. The chapter begins with a brief summary of literature related to change agents, intrapreneurship and institutional change. Following an outline of the research methodology the empirical data are presented and analysed.

From change agent to corporate entrepreneur

Kurt Lewin (1947) is attributed with initiating the terms 'change agent' and 'action research' in his quest to unite scientific inquiry with democratic methods. The first definition is credited to Lippitt *et al.* (1958: 10): 'The planned change that originates in a decision to make a deliberate effort to improve the system and to obtain the help of an outside agent in making this improvement. We call this outside agent a change agent.' Bennis (1964: 306) advanced the concept of change agents by defining them as professionals who, on the whole, 'have been trained and hold doctorates in the behavioural sciences' (see also Bennis *et al.* 1969). According to Beckhard (1969: 101) change agent refers to 'people, either inside or outside the organization, who are providing technical, specialist or consulting assistance in the management of a change effort'. Writers on organizational change generally present different interpretations of the term: some describe change agents as those who aid others 'with their *processes* of problem-solving and change, without themselves becoming involved in its content' (Dale 1974: 102). Rogers and Shoemaker (1971: 35) provided the following definition: 'a change agent is a professional who influences innovation decisions in a direction deemed desirable by a change agency.' One of the most influential views is that a change agent's 'primary role is to deliberately intervene in social systems' to facilitate social change (Tichy 1975: 772). Havelock and Havelock (1973: 60) argue that there are four distinct roles which change agents adopt: catalyst, solution giver, process helper or resource linker. Initiators and facilitators of change whether they act alone or as a part of a team may be grouped into four categories: external pressure groups, internal pressure groups, organizational development consultants and internal change agents or consultants (Duncan 1978). Others such as Beer (1980) use the terms 'manager', 'change agent' and 'consultant' interchangeably in referring to individuals responsible for generating and leading organizational change, while Kanter (1983) whose work is widely quoted uses the terms 'changemaster' and 'corporate entrepreneur' interchangeably.

It is not always clear whether internal or external change agents need to identify key stakeholders and accept 'ownership of change' (Leigh and Walters 1998). Internal change agents do have advantages in that they are familiar with systems and know where power lies, and they understand norms within the system (Greiner and Schein 1988). Some argue that CEOs should play a leading role in encouraging innovation and change (Stein and Pinchot 1998). Others such as Bird (1992) provide practical guides to effective change leadership and change (also see Rye 1996; Hussey 1998; Levesque 1999). Hartley *et al.* (1997: 62) state that literature on change agents 'has tended to produce idealized lists of skills rather than detailed studies of the actual roles, activities and performance of change agents in practice'. It is also evident that the majority of the change agent literature focuses on external consultants, particularly when action research is employed (Eden and Huxham 1996).

The terms 'intrapreneur' (Pinchot 1985) or 'corporate entrepreneur' (Burgelman 1983) are currently used more widely than change agent. Both entrepreneurship and intrapreneurship are often associated with social deviance. Hagen (1962), for example, argued that entrepreneurs emerge in communities that have been displaced or have experienced 'status withdrawal'. In the UK, Quakers, who were excluded from the professions, became entrepreneurs and founded their own businesses (Prior and Kirby 1998). Similarly, the intrapreneurship literature features many examples of social deviants who disobey superiors by continuing to work on projects to which they were committed (see Pinchot 1985). A typical case is provided by Digital in the 1980s when the company established Project PRISM, led by Dan Dobberpuhl, to develop a RISC-based microprocessor (Reduced Instruction Set Computer) for a new range of PCs (Katz 1994). Some senior managers were not convinced of the need for RISC-based chips because the company's existing VAX-based computers were still selling well. The outcome of managerial ambivalence towards Project PRISM was that the team were constantly halted or redirected. By 1987 the new technology had been adopted widely outside Digital and senior management decided to buy in a RISC-based chip rather than wait for Dobberpuhl's team to finalize PRISM. This was a massive disappointment to the project team particularly as PRISM was technically superior to the bought-in chip. Dobberpuhl allowed his team to spend time finalizing and testing the RISC design even though the project had been cancelled. After four months, PRISM was completed and operating speeds were three times faster than its rival. Dobberpuhl made a presentation to Digital's strategy group and despite some scepticism gained approval for the project to be reinstated. Eventually, the PRISM chip became a key source of the company's competitive advantage (Katz 1994).

Following a similar theme, Shane (1994) suggests corporate entrepreneurs must take on championing roles related to hierarchy, autonomy, equality, persuasion, monitoring and cross-functional ties. Differences between champions and non-champions based on the six roles were evaluated by questionnaire data from 4405 respondents representing forty-three organizations in sixty-eight countries (Shane 1994). Champions were defined as those who had overcome obstacles to innovation by taking action, involving some personal risk, outside their formal job description. According to the results champions were more likely to bypass hierarchy, preferred autonomy, encouraged equality among subordinates, and were less likely to demand financial justification for innovation. There were no differences according to team monitoring as both groups preferred teams to obtain authority before proceeding to the next stage of the innovation process. Differences between groups were significant for cross-functional ties but not in the expected direction. Non-champions preferred to establish cross-functional ties by obtaining support from other departments. Perhaps more significantly, deviant behaviour appeared to 'transcend culture' because in all sixty-eight countries 'champions hold significantly different views from non-champions about five different roles for champions in established organizations' (Shane 1994: 417).

Sharma and Christian (1999: 11) suggest there is a 'striking lack of consistency' in definitions of both entrepreneurship and intrapreneurship. A range of terms are adopted to describe what is broadly the same phenomenon and these include corporate entrepreneurship, corporate venturing intrapreneurship and corporate renewal. Some authors associate corporate entrepreneurship with business diversification through the development of new ventures, products or markets (Burgelman 1983; Zahra 1996). In contrast, Pinchot (1985) suggests that intrapreneurs are 'dreamers' who take responsibility for innovation of any kind within organizations. The definition which fits the perspective here comes from Chung and Gibbons (1997: 14): 'Corporate entrepreneurship is an organizational process for transforming individual ideas into collective actions through the management of uncertainties.'

According to Covin and Slevin (1991) 'independent' and 'corporate' entrepreneurs (see Collins and Moore 1970) share three postures: risk-taking, innovativeness and proactiveness. An alternative view is put forward by Covin and Miles (1999) who argue that in some cases organizations themselves adopt an entrepreneurial philosophy (Stevenson and Jarillo 1990; Stopford and Baden-Fuller 1994). The authors draw on work by Lumpkin and Dess (1996) who set out five dimensions of organizations which adopt an entrepreneurial orientation: autonomy, innovativeness, risk-taking, proactive and competitive aggressiveness, although Lumpkin and Dess (1996) are equivocal about whether all five elements will be present in entrepreneurial firms. Covin and Miles (1999) suggest that there are two elements which define entrepreneurial organizations. First, innovation, the introduction of a new product, process, technology, system, technique, resource or capability, is 'at the centre of the nomological network that encompasses the construct of corporate entrepreneurship' (Covin and Miles 1999: 49). The second element is that of sustaining high performance or radically improving competitive standing:

> Corporate entrepreneurship is engaged to increase competitiveness through efforts aimed at the rejuvenation, renewal and redefinition of organizations, their markets or industries. . . . It is the spark and catalyst that is intended to place firms on a path to competitive superiority or keep them in competitively advantageous positions.
>
> (Covin and Miles 1999: 50)

In his seminal contribution to the literature, Drucker (1994) identifies a number of US-based companies which demonstrate long-term commitment to corporate entrepreneurship; these include 3M, Johnson and Johnson and General Electric. Transforming mature firms into entrepreneurial organizations requires a management team that is visionary, persistent and willing to allow individuals to adopt entrepreneurial roles. Corporate entrepreneurs must extend existing capabilities without breaking links with the organization's core competences (Floyd and Wooldridge 1999).The authors go on to argue that middle managers are the locus of corporate entrepreneurship because they are central to the resolution of the capability–rigidity paradox (Leonard-Barton 1994). In developing their conceptual framework Floyd and Wooldridge (1999) integrate concepts from two distinct literatures. Knowledge theory emphasizes the importance of subjectivism, empiricism and pragmatism as central to the validation of organizational beliefs (Spender 1996). Social network theory provides insights into the role of weak ties, actor centrality and emergent networks as a basis for opportunity recognition (Granovetter 1973; Galaskiewicz and Wasserman 1993).

> Combining the knowledge and social elements, the model suggests that opportunities for entrepreneurship are perceived within organizations because individuals have access to unique information through weak social ties and because they are willing to accept ideas based on subjective criteria.
>
> (Floyd and Wooldridge 1999: 133)

Hitt and Ireland (2000) have been at the forefront of attempts to integrate corporate entrepreneurship with strategic management. They claim that, although the study of entrepreneurship remains underdeveloped in comparison to strategic management's relative maturity, there are six points of 'intersection': innovation, organizational networks, internationalization, organizational learning, top management teams and governance, growth and flexibility. Furthermore, it is suggested that there is 'convergence' in research related to the two topics with a focus on longitudinal design, dynamic analytical methods, structural equation modelling and cognitive mapping. At the same time, there is acceptance that there is a role for 'systematic qualitative research' based on ethnography, case surveys and multi-case methods (Hitt and Ireland 2000). Perhaps the most widely known author in the area of corporate entrepreneurship is Kanter (2000) who dismisses the idea that innovation is chaotic and random, insisting that 'structure and social arrangements' can actively stimulate innovatory activity. She claims that 'at its very root, the entrepreneurial process of innovation and change is at odds with the administrative process of ensuring repetitions of the past' (Kanter 2000: 168).

Kanter stresses the importance of integrative cultures in providing support for innovation and entrepreneurship. In addition, she discusses a range of factors that are essential if entrepreneurship is to flourish including 'power tools', which comprise three basic commodities:

1 information (data, knowledge, political intelligence and expertise);
2 resources (funds, materials, time and space);
3 support (endorsement, backing, approval and legitimacy).

While Kanter's work is important in understanding corporate entrepreneurship her writing is concerned entirely with the creation of innovatory cultures in large organizations. A recent study of twenty-three manufacturing-based small and medium-sized enterprises (SMEs) (average eighteen employees per firm) used constructs from the literature believed to influence innovation-supportive cultures (Chandler *et al.* 2000). Statistical analysis of employee responses provided a four-factor solution: innovatory supportive culture (ten items), organizational reward systems (seven items), management support (five items) and workload pressure (six items). The instrument also measured employee perceptions of managerial practices: management involvement, closeness to customer, waste, quality tools, clean-up. CEOs provided information on three constructs: competitive intensity, industry dynamism and environmental munificence. Finally, human resource management practices were examined: incentive schemes, formalized job descriptions, formalized appraisals, employee training programmes, top management involvement in recruiting and training, consistent disciplinary procedures, extensive induction procedures and formalized programmes for employee participation. The results indicated that creating innovation-supportive cultures requires that 'employees perceive management support and organizational support systems are consistent with a commitment to innovation on the part of management' (Chandler *et al.* 2000: 61). Excessive workload pressure was seen to inhibit innovation, and formalized HRM policies also had a negative impact on employee perceptions of an innovation-supportive culture. The authors point out that this is consistent with earlier work (Harrison 1987; Welbourne and Andrews 1996) which suggests that formal HRM practices appear to reduce opportunities for employee initiative and creativity. While the results provide some insights into the importance of innovation-supportive cultures, Chandler *et al.* (2000) acknowledge the limitations of their study, suggesting the need for 'larger and more diverse sample sizes' as well as 'fine grained' methodologies based on case studies.

Entrepreneurs innovate for themselves whereas intrapreneurs innovate on behalf of their employing organization. According to Carrier (1996), this difference in context is the key distinguishing feature. Carrier goes on to argue that research related to intrapreneurship falls into two categories: the first is concerned with individual psychological or personal attributes (Pinchot 1985; Carbone 1986), and the second focuses on the 'intrapreneurial process' by examining the organizational conditions which encourage such behaviours (Covin and Slevin 1991; Hornsby *et al.* 1993; Zahra and Pearce 1994). But Carrier (1996) notes the 'somewhat surprising' absence of research into intrapreneurship in smaller firms, particularly given that such organizations are ideal incubators. In her in-depth study of five small firms, Carrier (1996: 14) found that 'the structural and relational aspect is by far the most important factor in building an intrapreneurial environment'. In particular, the owner-manager's focus on growth, their strategic objectives as well as the types of reward, have a crucial impact on intrapreneurship. From the perspective of the intrapreneurs themselves, they were usually stimulated by challenges associated with innovation and learning but were also dissatisfied with the lack of extrinsic rewards in recognition of their efforts. Although there are similarities between the activities of intrapreneurs (corporate entrepreneurs) in small and large firms there are also significant differences. As Carrier (1996: 15) puts it: 'Small business entrepreneurship has its own special characteristics. The friendlier, more flexible structures of small businesses mean that the logic is one of convergence and matching rather than the identification of intrapreneurs.'

Institutional change

The concept of institution (Barley and Tolbert 1997) has returned to prominence in organization studies as a result of research by scholars such as Meyer and Rowan (1977), Zucker (1983) DiMaggio and Powell (1983) and DiMaggio (1991). This signifies a rejection of functionalist theories which portray efficiency as the driving force behind decision-making and adaptation to the environment as suggested by contingency theorists (Woodward 1958; Pugh and Hickson 1976). New institutionalists also reject the behaviourist aggregation of individuals by emphasizing cultural influences on both decisions and organizational structures because individuals operate 'in a web of values, norms, rules, beliefs and taken-for-granted assumptions that are at least partially of their own making' (Barley and Tolbert 1997: 93). In other words, institutions act as constraints on what actors can undertake but those constraints can be modified over time. Such a view of the links between individual and institution clearly has strong similarities with structuration theory (Giddens 1984). New institutionalism has been criticized recently for failing to account for the role of individual actors (Fligstein 1997; Beckert 1999; Hasselbladh and Kallinikos 2000). If institutional theory is to advance then there must be more explicit acknowledgement of the interdependence of actions and institutions (Barley and Tolbert 1997: 94). This may be done by developing a more dynamic model of institutions (Whittington 1992) as well as by adopting methodologies which investigate recursive links between action and institutions. More recently, Beckert (1999: 778) argues that although institutional theory is powerful in demonstrating the way in which organizations are linked to their environment the role of agency is underestimated. It is therefore important to examine the processes by which strategic choice (Child 1972) is exercised within organizations. Textbooks (and lecturers) emphasize the managerial role in most organizational activity ranging from strategic business planning to day-to-day decision-making. In general, new institutionalists pay little attention to links between individual agency and institutional change. As Beckert (1999: 778) puts it: 'If, however, we assume that in many situations agents "make a difference", it becomes a weakness of institutional theories that they cannot account for the role of strategic agency in the processes of organizational development.'

Some institutional theorists acknowledge that conflictual and contradictory rules can encourage discretionary behaviour (DiMaggio 1988; DiMaggio and Powell 1991; Scott 1991, 1994, 1995; Kondra and Hinings 1998). There has also been some focus on the role of institutional entrepreneurs who access scarce organizational resources (Fligstein 1997). In linking entrepreneurship with institutionalist approaches, Beckert (1999) draws on the Schumpetarian distinction between managers who act according to organizational routines, and entrepreneurs who are innovators concerned with changing existing routines or instigating new ones. Entrepreneurs are seen as creators of strategic opportunities while managers focus on stability and embeddedness (institutional rules). According to Giddens (1984: 64), the human need for 'ontological security' means that actors 'stick to routine patterns of behaviour that unintentionally reproduce the structures of their worlds'. This has important implications for the study of innovation and entrepreneurship because of the suggestion that actors remain wedded to activities with which they are most familiar. The degree of innovation, incremental or radical, can lead to a modification or replacement of rules and resources. Innovation is subversive because it undermines existing practices, and constitutive because it creates new practices and routines (Schumpeter 1934; Nelson and Winter 1982; Orlikowski 1992). According to Beckert (1999: 789): 'Action cannot be understood as the simple execution of existing scripts, but develops in a duality between agency and structure.'

Links between actors and institutions is similar to that between grammar and speech: 'every expression must conform to an underlying set of tacitly understood rules that specify relations between classes of lexemes' (Barley and Tolbert 1997: 96). In structuration theory the 'institutional realm' is linked to the 'realm of action' by modalities described as interpretative schemes, facilities and norms (Giddens 1984). Barley and Tolbert (1997: 98) suggest that 'scripts' (observable, recurrent activities and patterns of interaction characteristic of a particular setting) may be used in place of modalities because they can be empirically identified more easily than the abstract notion of modalities. Institutionalization occurs as actors unconsciously enact scripts which encode institutional principles (cf. Nelson and Winter 1982) and eventually behaviour patterns are objectified into organizational norms (Barley and Tolbert 1997: 102). Such a conceptualization of the links between action and institution means that researchers must record observational data on patterns of interaction and ways in which actors interpret their own behaviours. The latter aspect is important because it is possible to evaluate whether actors considered alternative courses of action. In addition to direct observation, interviews and questionnaires, Barley and Tolbert (1997: 105) suggest that researchers should make use of various forms of archival data including historical documents retained by organizations as well as personal diaries. They point out that 'institutional researchers have made little use of such data, perhaps in part, because they are unaware of its existence and because most sociologists are not schooled in the ways of historians'.

Incorporating managerial agency

To examine their hypothesis that strategic leadership is particularly important during periods of rapid change Denis *et al.* (1996) studied two hospitals in Quebec. They analysed the dynamics of change under ambiguity by collecting documentary records, sixteen retrospective interviews and 'ethnographic' data gathered by one author who worked as a consultant in a community health department attached to one of the hospitals. The authors draw on the work of Hinings and Greenwood (1988) in arguing that an organization's structural characteristics are supported by 'interpretative schemes' which in turn are based on collective ideas, beliefs and values. The 'leadership role constellation' and 'influence tactics' are conceptualized as the two key dimensions of the change process which link interpretative schemes and organizational structure. Leadership is seen as a team rather than an individual phenomenon (Pettigrew 1992) and constellation effectiveness is influenced by three factors: specialization, which refers to the extent to which roles have narrow areas of expertise; differentiation, i.e. the division of roles to ensure that there is not too much 'overlap'; complementarity and the interlocking of roles (Denis *et al.* 1996: 693). Organization influence tactics (Pfeffer 1981) describes the way in which individuals and groups attempt to influence the course of events by selective use of criteria such as agenda-setting, coalition-building, cooptation and symbolic management (Denis *et al.* 1996: 693). Tactics have three different outcomes. First, there are symbolic changes based on modifications to existing interpretative schemes. Second, there are substantive changes based on organizational restructuring. Third, there is political change based on formal–informal power relations and the evolution of leadership roles. The case analysed by Denis *et al.* (1996: 689) 'illustrates the essentially collective and fragile nature of leadership in ambiguous organization'.

Day (1994) carried out a questionnaire-based survey of 136 internal corporate ventures to examine the role of the 'principal champion'. She distinguishes between bottom-up champions who are close to the technical or market interface and top-down champions who are particularly important in times of major changes because they initiate substantive and symbolic actions. A third category, the dual-role champion, combines the role of product champion (bottom-up) and organizational sponsor (top-down) through their ability to mobilize knowledge, information and power. The study, based on a traditional positivistic methodology, adopted innovation (timing, life cycle, technological newness) as the dependent variable. The independent variables included the principal champion (product champion or organizational sponsor), hierarchical level and organization location (proximity to core). A number of control variables such as size, age, diversity, R&D spend, market size, 'cannibalism', location and supporting assets were also included. The results indicate that organizational size did not have an impact on innovativeness but provided 'weak support for the argument that firms ossify as they age' (Day 1994: 164). R&D spend was positively related to innovation and diversity had a negative relationship with the introduction of new products. This is because, Day surmises, there is less incentive to innovate or to enter new markets, while the overall level of explanation, with an R^2 of 0.25 significant at the 0.001 level, means that although 'much is left unexplained, these results are very promising' (Day 1994: 164). Dual role champions were found to be more common (36 per cent) than either bottom-up (30 per cent) or top-down (20 per cent) champions.

Following a Giddensian approach and drawing on key elements of the above literature my conceptualization of the links between corporate entrepreneurship and institutional change is as follows. First, corporate entrepreneurs create strategic opportunities by instigating change in existing organizational routines (Schumpeter 1934; Beckert 1999). Second, corporate entrepreneurship is not based on the activities of a single individual but it is essentially a collective effort which demands high levels of collaboration to achieve successful transformation. Managerial actions to initiate change can have three different outcomes: symbolic, substantive and political change (Denis *et al.* 1996). Third, corporate entrepreneurs can adopt one of three roles: top-down champion, bottom-up champion or dual-role champion. This illustrates the importance of recognizing that even though change must be a team effort there will generally be one individual who acts as the champion in taking on responsibility for building and motivating 'collective transformational leadership' (Day 1994). Fourth, to be effective, corporate entrepreneurs must adopt a championing role which involves the breaking of organizational rules (Shane 1994).

Research methods

In stating the advantages of case study research, Yin (1994) claims that observing a 'chronological sequence' permits investigators to 'determine causal events over time'. My view is that establishing causality in highly complex social organizations is extremely difficult whatever methodology is adopted. Rather, I concur with Barley (1986: 81) who argues that mapping 'emergent patterns of action demands a detailed qualitative approach'. He continues: 'Retrospective accounts and archival data are insufficient for these purposes since individuals rarely remember, and organizations rarely record, how behaviours and interpretations stabilize over the course of the structuring process.' Longitudinal research remains rare in organizational studies and single cases raise issues of generalizability. In discussing the shift from micro to macro levels Hamel *et al.* (1993) argue that the objectives are more important than the number of confirmatory cases. This refers to the distinction between statistical generalization (Yin 1994), in which inference is made about a specific population, and analytical generalization, in which empirical data are compared with a theoretical 'template'. In this study I adopt an approach based on structuration theory which is described as 'a process-oriented theory that treats structure (institutions) as both a product of and a constraint on human action' (Giddens 1984: 2). Such a methodology helps bridge the determinism associated with structural accounts and the voluntarism of social action.

Data are drawn from a study of MFD, a privately owned manufacturing company founded over forty years ago to supply casting and machined components to the Ministry of Defence. Until recently, MFD manufactured to contract and had little marketing expertise. As one MFD manager stated: 'if you took the customers away you would have difficulty identifying production of a specific product.' One interesting decision was turning down a contract to manufacture Dyson's dual cyclone vacuum cleaner. As the manager went on to say, 'the opportunity didn't fit the current profile of the business but Dyson also expected MFD to do the marketing'. By the late 1990s there was increasing emphasis on higher volume electronic assemblies for two major customers: BT and LaComm.

Access to MFD was negotiated originally when senior managers agreed to participate in a doctoral research project investigating the nature of innovation networks in a range of mature manufacturing firms (Beckinsale 2001). As supervisor, I visited the company and realized that it fitted my own research interests in the role of corporate entrepreneurship. Data were acquired from a variety of sources including observation, regular discussions with GW (see below), company documents as well as fifteen interviews carried out over a two-month period at the end of 2000. These semi-structured interviews, taking between forty-five and sixty minutes, were taped, and recorded recent managerial views on events within MFD. Interviewees generally confirmed that change was initiated by the owner and managing director (MF) who was concerned about declining activity within the company as a result of fewer defence contracts. At an operational level, the change agent was GW, a middle manager who, in contrast to most other white- and blue-collar workers, had joined the company less than one year before the change programme began.

Kicking against the pricks

The issue of corporate entrepreneurship is examined by investigating the interaction of institutional structures and individual agency during a period of major change within MFD. The study is influenced broadly by literature associated with intrapreneurship and corporate entrepreneurship in which it is generally assumed that such individuals must be prepared to adopt a championing role which involves challenging or subverting organizational rules (Shane 1994). In other words, corporate entrepreneurs must be prepared to 'kick against the pricks' represented not only by rules and routines but also by those unwilling to change or who have a vested interest in maintaining the status quo. Activities within MFD are analysed by means of four factors discussed on pp. 130–131.

Strategic opportunities and changing routines

Within MFD two significant events which occurred in 1996 to 1967 seem to have stimulated a re-evaluation of the company's strategic opportunities: first, recognition throughout the management team that they had to pay much greater attention to customers, and second, a willingness by MF, the owner, to invest considerable sums of money in new equipment. GW instigated an investment of £250,000 in an IT system to integrate a wide range of activities including materials management and payroll. The decline in UK defence spending stimulated change because until 1990 MoD contracts accounted for 60 per cent of MFD's business. The firm was heavily reliant on MF's personal network of contacts in the military which he often visited in his private helicopter to generate new contracts. Significant reductions in defence spending and the move away from cost-plus contracts to a market-based approach meant that by 1998 MFD had lost all MoD work.

> It was first obvious with the recognition that the defence business would not be enough for us to survive on. The family have now recognized that there are other ways of doing things but that is not the only answer. I think that there are other people who are going to determine how the company will be run rather than the familiar immediate group.
>
> (Materials Director)

Survival meant that new products and new customers were a necessity. Therefore, change in MFD was a reaction to external threats rather than a proactive search for new opportunities. MF initiated a number of changes including the recruitment of managers with broader industrial experience, including GW who had spent more than twenty years working for a large domestic appliance manufacturer organized according to Fordist principles. His ideas on material flows and the elimination of WIP were revolutionary to most long-serving MFD managers. A new senior marketing position was filled by PD who had vast experience in a range of industries, including time spent as MD of a medium-sized manufacturing company. Purchasing the IT system also led to the employment of a small team with high-level technical skills in both software and hardware. As discussed below, the two senior appointments had symbolic, substantive and political significance. These developments led to considerable changes to the routines (systems) with particular emphasis on greater openness, trust, flexibility and interaction with customers (there was broad agreement across the managerial group so illustrative quotations are representative):

> It takes a lot of commitment from the guy signing the cheque. He can't have cold feet at the first hiccup, he has to be 100 per cent behind it and ask what will it take to do it, not can we do it, should we do it? Those are questions of the past. We've answered those and now it's time to ask what it will take to do it? Don't tell me the problems tell me if I can help you solve them. This is new to MFD, total commitment from top-down and from bottom-up.
>
> (Telecom Manager)

> I think the main change has been in a greater flexibility in terms of decision-making and authority, more flexibility in terms of freedom of movement, in terms of what we can and can't do. Opening up opportunities, you could call it greater trust. This in turn reflects on the people working for me as well, I endeavour to give them more flexibility. In the time I have been with the company, probably in the last two or three years I have seen more change than the rest of the time. I'm not sure whether that was a conscious decision made elsewhere or if it is just the way that we have had to operate in terms of speed and response.
>
> (Purchasing Manager)

Evaluation of the changed external environment was primarily the responsibility of the owner and, as described above, it became obvious that reliance on MoD contracts was at an end. Recruitment of new staff was the responsibility of KC (works director) as MF did not get involved in operational activities. There was gradual acceptance that the traditional top-down approach was no longer appropriate and that employees at all levels had to be given greater responsibility for their day-to-day tasks. A further significant structural change was the creation of 'module champions' who were given responsibility for liaising between their departments and the team introducing the new computer system.

> It's difficult to be certain about the 'relaxation' but there has been a relaxation and I am not exactly sure about the motives behind that but hopefully it is a greater trust in the people further down the line.
>
> (Purchasing Manager)

> In one way it is quite dramatic, it has caused the company to look critically at the business and take mostly appropriate action. LaComm is a close-knit team and that side is now considerably stronger. Also the company is confident to take that part of the business forward because it has a better understanding of what it is good at and therefore what it can market.
>
> (Marketing Director)

> We've gone in at the bottom level and said we have to improve the response time to customers, improve the service [and] we need to do it in a number of ways. They [operators] feel that they have more responsibility and they have actually thought about us as people rather than numbers. It has raised spirits and that has helped enormously with the success of the project.
>
> (Industrial Engineer)

Symbolic, substantive and political change

According to Denis *et al.* (1996) effective change needs to be managed through high levels of collaboration. Further, the effect of strategic leadership can be categorized according to three outcomes: symbolic, substantive and political change. There was evidence of all three factors in MFD. Symbolic change occurred as a result of MF's willingness to spend more time in the plant talking to supervisors and managers. His earlier remoteness was attributed partly to difficulty in relating to people and the more open approach was a catalyst for broader changes which encouraged greater trust between managers and shop-floor workers. Other symbolic changes emphasized the increasing importance of customers:

> I think that we've had it very cosy and the real world is starting to hit us. LaComm have told us that we've got to find more customers because they're competing against slick, lean operations and they can't support us. We've got to go out and win business against other companies. We'll have to prove we're committed.
>
> (QA Manager)

> The company has always manufactured to customer requirements but that is a reactionary position. A big impact is that now we are proactive and draw customers in. That is a dramatic difference and the awakening of that reality was brought about by LaComm but acceptance from the chairman down was not that easy.
>
> (Materials Director)

Substantive change occurred as a result of two major investments, the first of which was the IT system approved in 1998. As the second, in December 1999 GW instigated a major capital investment in the assembly area. More than £350,000 was spent on a process line for automatically assembling printed circuit boards and a major reorganization to give a logical work flow and remove excessive work-in-progress. This was partly to satisfy the demands of LaComm which wanted MFD to adopt a more professional approach that would impress its own customers who sometimes visited subcontractors.

> Mr F (MD) has spent a lot of money during the last two or three years. First of all if he hadn't we would be out of business because LaComm and BT would go elsewhere even if it was only to second-source suppliers. We're tooled up for the electronics trade and we need to stay in it. We're buying dollops of equipment – a third of a million pounds a time.
>
> (Works Manager)

> We've made significant improvements in quality standards, all round, image, housekeeping, general labour efficiency are all dramatically up (50 per cent). All of that is very good stuff and as well as that £100,000 is being spent on equipment. All of this originates from proper capacity planning and preparation. We have an awful lot to do before we are a really serious class act, but we're moving in the right direction.
>
> (Production Supervisor)

Political change occurred with a shift from a traditional authoritarian management style to an approach that had elements of consultation if not participation. The works director's (KC) wife was personnel manager until 1995 and had an extremely authoritarian approach. On retirement, her replacement immediately instigated a more conciliatory and democratic approach to relations with shop-floor workers. BG who acted as materials manager and production manager in the assembly area was also extremely authoritarian. His office overlooked the shop-floor and if he saw operators talking he would immediately summon the supervisor and demand that they be disciplined. The supervisor described his tactics for dealing with these situations:

> I would go up to them and start waving my arms about as if I was giving them a right bollocking – but I would probably be saying 'come on boys get back to work that bastard is watching you from the office'. I decided if I ever became a manager I would behave very differently – he taught me what not to do.

Early in 1999, GW was given managerial responsibility for the assembly areas and BG reverted to his previous role of materials manager. GW gave supervisors much greater responsibility for dealing with shop-floor issues while he focused on output and quality. Greater participation in decision-making at all levels and more consultation with shop-floor workers was in evidence throughout the organization.

> Working practices have changed we're in the modern world now. Everything has changed so dramatically from purchasing to manufacturing with the help of computers. The whole philosophy of the firm has changed and it seems to have gone in a different direction, the right way I think. Now many more people get a say in what goes on. There are little subcommittees and everyone is involved.
>
> (Stores Manager)

> I've been involved in the change in working practices, the change in the type of people that we are employing – different skills, retraining of employees, changes in health and safety requirements and all the associated things that go with it, costs, manual handling, risk assessments all the rest of it.
>
> (Personnel Manager)

> We've had successes there is no doubt about that in the sense that we've built a team environment rather than a tiered managerial environment on the telecoms side. That has been driven by the introduction of new blood and by recognizing the potential of some we already had in the business. We've been able to form the foundation of a much bigger business.
>
> (Materials Director)

Corporate entrepreneurship

GW was given charge initially of management services with primary responsibility for the control of labour costs via the issue of standard times. He quickly realized that there were a wide range of factors influencing the inefficient use of labour including an ancient and inflexible MRP system which made it extremely difficult to track flows of material through the factory. This was crucial because as a result of material shortages operator 'waiting time', paid at average earnings, was high. The work of white-collar staff was also inefficient, as supervisors, foremen and storekeepers spent a considerable amount of time searching for missing materials. After carrying out a detailed analysis GW decided that the only way to improve efficiency was to purchase a new mainframe computer with software, including MRPII, capable of dealing effectively with the complexity of operations within MFD. GW presented the results of his analysis and recommendations to KC with an estimate of the total cost of the project including the employment of two-thirds of new technical staff. Following discussions with MF, KC approved the project and GW and his department were given responsibility for the purchase, installation and commissioning of the new system.

The computer was intended primarily to resolve problems associated with stock control and labour inefficiency but it was also designed to link all major functions within MFD. This meant obtaining departmental heads' co-operation to improve transparency and to incorporate their requirements into the specification. One of the most intriguing aspects of MFD was the management style of KC (works director) who, to the frustration of department heads, never held management meetings. Rather, KC's approach was to discuss such changes on an informal basis obtaining the views of each individual manager. This provided GW with the opportunity to take the initiative in organizing management meetings to discuss implementation of the new system in which he was able to outline his view of the company's future direction. In Day's (1994) terms, GW typified the dual-role champion by acting as corporate sponsor for two major capital investment projects and at the same time initiating numerous minor changes on the shop-floor. These changes included reorganization of material flows as well as allowing first-line supervisors to take on responsibility for day-to-day labour management problems.

> The biggest change for me has been the introduction of GW as our new manufacturing manager. We've invested in new technology in the last few years and with that came the initial learning curve which we've done very well. Now we have bought more equipment. The technicians get sent on courses on calibration and first-level maintenance.
>
> (Production Supervisor)

Early in 1999, when the mainframe system had been implemented and most 'teething problems' had been overcome, KC decided that GW's role should be extended to include responsibility for electronics manufacturing. Most managers identified GW as the key prime-mover in stimulating change in MFD and a number of quotes illustrate his success in becoming dual-role champion.

> The problem on the shop-floor has always been lack of parts that is the biggest moan that you will hear out there. But that is one of the reasons that GW has taken over telecoms, the guy who was doing that job was a

> director. Whenever products weren't being produced it was him not ordering parts but he didn't get a bollocking. Now it's better because GW shouts at KC to get things done. That's a success, putting GW out there.
>
> (IT Manager)

> Well it is very interesting because as you know I joined the company two and a half years ago. They were a very staid company, set in their ways and GW has introduced new methods and a whole different outlook and they seem to have reacted and it is very, very noticeable. They've realized that they have got to get into gear to keep up with industry.
>
> (BT Production Manager)

> The changes have been forced upon us, Y2K, we have to keep moving. They {management} don't seem fussed about how much we spend, although a little bit cautious at times. As long as it looks like we are doing something that will improve the company, improve business. GW has been good for MFD, he understands, he wants to push things forward.
>
> (IT Manager)

Breaking the rules

MF was the initiator and GW the corporate entrepreneur who took responsibility for introducing new equipment and new ways of working. He recognized that the company had to change but also realized that there was an opportunity for him to take the initiative.

> There has been an injection of new thinking. But that is more of a consequence than a catalyst and I still think that the reason for change is necessity. It's so much easier when the top people feel that something has to change even if they don't know what to do. This isn't a very bureaucratic business and it's very nice when you get even a hint of direction and it doesn't have to be rubber-stamped.
>
> (GW)

GW instigated change by gradually extending his responsibilities as he gained a fuller understanding of the way in which MFD operated and by identifying the main obstacles to a more effective business. GW was not openly confronted by middle and lower managers who did not share his vision of a more efficient and responsive organization. Although, in a reflection of his conviction that he had the appropriate solutions, GW was ruthless in encouraging the more reactionary individuals to leave or in ensuing that they were marginalized.

The most obvious examples of rule-breaking behaviour were associated with GW's analysis of MFD's failing related to the MRP system which was extremely ineffective in tracking material flows through the factory. Although this was not GW's area of responsibility his 'championing' of the new IT system was a catalyst for a number of other related changes. Up-to-date shop-floor information enabled the works manager and the materials manager to regain control of scheduling. As discussed in the previous section, the lack of regular management meetings allowed GW to create a forum in which all those whose jobs were affected by the new system were involved in discussion about its implementation. These meetings led to the creation of 'module champions' who were responsible for ensuring that the views of every department were incorporated into the system design. Uniquely for MFD, information about the selection, purchase and implementation of the new system was passed on via regular meetings over a twelve-month period. As a consequence, staff at all levels were given the opportunity to 'buy in' to the new way of working without having an 'alien' system imposed from above. There were occasions when GW was acting at the limits of his competence (Kanter 1983) as he took on responsibility for a number of activities with which he personally was unfamiliar including purchase of the MRPII system.

> I had no knowledge at all about computers, I'd only been here a few months but KC believed that I was capable of organizing, structuring, planning the putting together of the implementation, that is one example. It is not my area of expertise, but I did a satisfactory job.

There were high levels of deference to authority among the majority of MFD staff which reflected the strong paternalistic culture which had been created by MF. GW's willingness to cross both hierarchical and departmental boundaries helped encourage greater sharing of information within the company. In addition, his willingness to ensure that there was widespread consultation about the exact operational requirements mobilized the skills and knowledge of first-line supervisors and technical staff:

> GW has created a lot of movement, you have to give him credit for getting through to the top level as much as he has done. It was a very high wall to get over but he managed it and to be fair they have given him backup. He is very enthusiastic and that enthusiasm has transferred down.
>
> (QA Manager)

Institutionalizing change

In this section I summarize the changes which took place in MFD by implementing structuration theory. As discussed above, Giddens (1984) argues that three modalities – interpretative schemes, facilities and norms – link individual action (agency) to the institutional realm (structure). Therefore, I use these modalities, summarized in Table 7.1, to explain the way in which corporate entrepreneurship contributed to a restructuring of relationships within MFD. Three key interpretative schemes appear to have underpinned activities within the company before the research began. First, existing customers such as the MoD provided regular business and there was no emphasis on marketing activities. When the research began in 1997 transition was already underway as MoD business was in rapid decline and contracts had been obtained with BT and LaComm. Second, shop-floor activities were still organized according to the principles of low-volume batch manufacturing involving machining and simple assembly work. Third, both first-line supervisors and senior mangers were committed to maintaining 'arrears' as a 'guarantee' for shop-floor work. That is, by ensuring that work was carried out two, three, four or even more weeks behind schedule everyone knew that shop-floor activity could be maintained at least until arrears were cleared. This way of thinking had been encouraged by the 'cosy' relationship between suppliers and the MoD who seemed not to expect deliveries on time. Even when new business was obtained with customers such as BT the institutional 'routine' of working in arrears remained in place. Representatives of LaComm were unwilling to contemplate late deliveries, and the demands of their engineers and buyers was a catalyst for change. The problem within MFD was that, until the appointment of GW, there was no one within the company who had experience of efficient scheduling and mass production.

Power over other actors creates structures of domination in which the modality of facilities has two dimensions: allocative and authoritative resources (Giddens 1984). Command over allocative resources refers to those with power to make decisions about the purchase of goods, services and

Table 7.1 The structuration of organizational change

	Old modalities (pre-1997)	*New modalities*
Interpretative schemes	MoD provides steady business; batch manufacturing and low-tech; 'arrears' (WIP) guarantee work.	Seek and cultivate new customers; mass production and high-tech; shift to 'lean' manufacturing.
Allocative resources	Initiated from the top; internally generated.	Initiated from below; internally generated.
Authoritative resources	Proprietal; authoritarian/paternalistic; top-down.	Proprietal; professional management; committees/team-working.
Norms	Deference to authority; knowledge internally generated; senior mangers initiate action.	Greater openness/democracy; create external networks; all managers sanctioned to act.

equipment. Before 1997, all major expenditure had been initiated by the owner or works director (KC). GW not only suggested that the investment was necessary; he was also given responsibility for managing both major projects. Although there was a shift in terms of the way in which resources were accessed there was no change in the source of those funds. As MFD is a family-owned business there is no legal requirement for reporting financial results. Therefore, for most managers in the company finances remain a 'black box' to which they have no access. Consequently, decisions about whether particular projects should be funded remained entirely at the discretion of MF.

Authoritative resources describe an individual's ability to organize and co-ordinate the activities of other social actors. Despite widespread change, proprietorial rights embodied in ownership of the company remained the main source of managerial legitimacy. As discussed earlier, GW gradually emerged as a 'dual-role' champion because he was intimately involved with day-to-day activities on the shop-floor while at the same time having direct access to the works director. So while the family-owned company remained 'proprietorial' in terms of the authoritative resources there was a gradual shift of emphasis from authoritarianism to a more professional approach to management. Once again, the influx of new middle managers with extensive experience in other businesses helped reorientate existing skills and knowledge within the company. A key element in this process was the democratization of information as illustrated by regular meetings to discuss the MRPII system. The symbolic and political implications of this new approach encouraged the adoption of more team-working for the resolution of organizational problems.

Norms are the broadly accepted conventions and rules which legitimate conduct within any organization and are likely to influence strongly the actions of every individual. Because MFD remains family owned and is a major employer in a relatively isolated community there is still a strong culture of deference to paternalistic authority. At the same time, as most of the above quotations illustrate, there is 'a complete new way of thinking' within the company which has encouraged greater trust and democracy. The introduction of 'outsiders' to key middle-management positions has influenced institutional change in a number of ways. First, there are obvious modifications to the managerial hierarchy to accommodate the newcomers. Second, there is the creation of new mechanisms for the dissemination of information such as regular management meetings, committees and team-working arrangements. Third, and perhaps most importantly, managers (GW/PD) have brought new ways of thinking which emphasize the importance of accessing knowledge and expertise from external sources. For example, when purchasing the MRPII system, rather than calling on a company with which MFD had previously done business, GW initiated an evaluation procedure in which ten IT companies were required to submit detailed proposals (technical specification and costs). GW also encouraged

KC to take advantage of a programme developed by the WDA (Welsh Development Agency) to improve manufacturing techniques in SMEs. The WDA project, which emphasized the importance of Kanban and shop-floor teams, in combination with the MRPII system, helped MFD shift towards the principles of lean manufacturing.

Conclusions

MFD, a mature manufacturing company, has faced a number of crises over the past ten years including the loss of its main customer, the MoD. I have attempted to analyse the process and institutionalization of change encompassed by the three-year period from January 1997 to early 2000. First, the owner MF recognized the need for a more proactive and aggressive approach to marketing as a result of losing MoD contracts. Second, the research began at the same time as GW was employed as management services manager in January 1997. After a relatively short period in which he evaluated the existing situation GW suggested to the works director a number of major changes including two large investments in capital equipment. As suggested by Tolbert and Zucker (1983), champions must first define organizational problems and then provide appropriate solutions. Experience of a mass-production environment helped GW identify a lack of professionalism in managing the production process. Initially, he demonstrated his abilities by implementing a number of changes designed to improve labour productivity. Gradually, he extended his influence by, for example, implementing a system to collect 'heavy metals' used in manufacturing processes which were being 'flushed' into the company's drains. Diluting contaminants demanded extremely large quantities of water which was increasingly expensive. A relatively minor investment of £10,000 allowed the company to make substantial savings in the cost of water as well as being able to sell the recovered heavy metals.

In adopting the role of corporate entrepreneur, GW did not act as a rule-breaker as defined within the intrapreneurship literature. Rather, his approach was more subtle and involved stretching the boundaries in a way which did not alienate his superiors but provided him with the freedom to act without having to constantly seek approval from senior managers. He also won the support of other middle managers and their subordinates by encouraging full participation in decisions about new equipment and new ways of working. At the same time, because the company future was threatened by the loss of its defence-related work there was, throughout the organization, a willingness to accept radical new ideas. Equally, because MFD remains a family-owned business, managing director MF was able to act without reference to other shareholders. In summary, the change process within the company represents an important example of the interaction of individual agency and institutional structures.

Note

1 *The Bible*, Acts IX (Saul said, 'Who art thou Lord?' And the Lord said, 'I am Jesus whom thou persecutest: it is hard for thee to kick against the pricks'); Nick Cave and the Bad Seeds, 'Kicking Against the Pricks').

References

Barley, S. (1986) 'Technology as an occasion for structuring: evidence from observation of CT scanners and the social order of radiology departments', *Administrative Science Quarterly* 31: 79–91.

Barley, S. and Tolbert, P. (1997) 'Institutionalization and structuration: studying the links between action and institution', *Organization Studies* 18, 1: 93–117.

Beckert, J. (1999) 'Agency, entrepreneurs and institutional change: the role of strategic choice and institutional practices in organizations', *Organization Studies* 20, 5: 777–800.

Beckhard, R. (1969) *Organizational Development: Strategies and Methods*, Reading, MA: Addison-Wesley.

Beckinsale, M. (2001) *Strategic Innovation Networks*, Unpublished Ph.D., Aston Business School.

Beer, M. (1980) *Organization Change and Development. A Systems View*, Santa Monica, CA: Goodyear.

Bennis, W.G. (1964) 'The change agents', in R.T. Golembiewski and A. Blumberg (eds) *Sensitivity Training and the Laboratory Approach,* Ithaca: Peacock.

Bennis, W.G., Benne, K.D. and Chin, R. (1969) *The Planning of Change* (2nd edn), New York: Holt, Rinehart & Winston.

Bird, M. (1992) *Effective Leadership: A Practical Guide to Leading Your Team to Success*, London: BBC Books.

Blau, P. (1974) *On the Nature of Organizations*, New York: Wiley.

Burgelman, R.A. (1983) 'Corporate entrepreneurship and strategic management: insights from a process study', *Management Science* 29: 1349–1364.

Carbone, T.C. (1986) 'The making of a maverick', *Management World* 15, 5: 31–33.

Carrier, C. (1996) 'Intrapreneurship in small businesses: an exploratory study', *Entrepreneurship: Theory and Practice* 21, 1: 5–21.

Chandler, G., Keller, C. and Lyon, D. (2000) 'Unravelling the determinants and consequences of an innovation-supportive organizational culture', *Entrepreneurship: Theory and Practice* 25, 1: 59–78.

Child, J. (1972) 'Organizational structure, environment and performance: the role of strategic choice', *Sociology* 6: 1–22.

Chung, L.H. and Gibbons, P.T. (1997) 'Corporate entrepreneurship: the roles of ideology and social capital', *Group and Organization Management* 22, 1: 10–30.

Collins, O. and Moore, D.G. (1970) *The Organization Makers*, New York: Appleton.

Covin, J. and Miles, M. (1999) 'Corporate entrepreneurship and the pursuit of competitive advantage, *Entrepreneurship: Theory and Practice* 23, 3: 47–62.

Covin, J. and Slevin, D.P. (1991) 'A conceptual model of entrepreneurship as firm behaviour', *Entrepreneurship: Theory and Practice* 16, 1: 7–25.

Dale, A. (1974) 'Coercive persuasion and the role of the change agent', *Interpersonal Development* 5, 2: 102–111.

Day, D.L. (1994) 'Raising radicals: different processes for championing innovative corporate ventures', *Organization Science* 5, 2: 148–172.

Denis, L., Langley, A. and Cazale, L. (1996) 'Leadership and strategic change under ambiguity', *Organization Studies* 17, 4: 673–699.

DiMaggio, P.J. (1988) 'Interest and agency in institutional theory', in L. Zucker (ed.) *Institutional Patterns and Organizations*, Cambridge: Ballinger.

DiMaggio, P.J. (1991) 'Constructing an organizational field as a professional project: US art museums, 1920–1940', in W.W. Powell and P. DiMaggio (eds) *The New Institutionalism in Organisation Analysis*, Chicago, IL: Chicago University Press.

DiMaggio, P.J. and Powell, W.W. (1983) 'The iron cage revisited: institutional Isomorphism and collective rationality in organizational fields', *American Sociological Review* 48: 147–160.

DiMaggio, P.J. and Powell, W.W. (1991) 'Introduction', in W.W. Powell and P.J. DiMaggio (eds) *The New Institutionalism in Organizational Analysis*, Chicago, IL: Chicago University Press.

Drucker, P. (1994) *Innovation and Entrepreneurship*, Oxford: Butterworth-Heinemann.

Duncan, W.J. (1978) *Organizational Behaviour*, Boston, MA: Houghton Mifflin.

Eden, C. and Huxham, C. (1996) 'Action research for the study of organizations', in S. Clegg, C. Hardy and W. Nord (eds) *Handbook of Organization Studies,* London: Sage.

Fligstein, N. (1997) 'Social skill and institutional theory', *American Behavioural Scientist* 40: 397–405.

Floyd, S. and Wooldridge, B. (1999) 'Knowledge creation and social networks in corporate entrepreneurship: the renewal of organizational capabilities', *Entrepreneurship: Theory and Practice* 23, 3: 123–145.

Galaskiewicz, J. and Wasserman, S. (1993) 'Social network analysis: concepts, methodology and directions for the nineties', *Sociological Methods and Research* 22, 3: 22–41.

Giddens, A. (1984) *Theory of Structuration*, Berkeley: University of California Press.

Granovetter, M. (1973) 'The strength of weak ties', *American Journal of Sociology* 78, 6: 1360–1380.

Greiner, L. and Schein, V. (1988) *Power and Organization Development: Mobilizing Power to Implement Change*, Reading, MA: Addison-Wesley.

Hagen, E.E. (1962) *On the Theory of Economic Change: How Economic Growth Begins*, Homewood, IL: Dorsey.

Hamel, J., Dufour, S. and Fortin, D. (1993) *Case Study Methods*, London: Sage.

Harrison, R. (1987) 'Harnessing personal energy: how companies can inspire employees', *Organizational Dynamics* 16, 2: 4–20.

Hartley, J., Benington, J. and Binns, P. (1997) 'Researching the roles of internal change agents in the management of organizational change', *British Journal of Management* 8: 61–73.

Hasselbladh, H. and Kallinikos, J. (2000) 'The project of rationalisation: a critique and reappraisal of neo-institutionalism in organization studies'. *Organization Studies* 21, 4: 697–720.

Havelock, R.G. and Havelock, M.C. (1973) *Training for Change Agents*, Michigan, OR: Institute for Social Research.

Hinings, R.C. and Greenwood, R. (1988) *The Dynamics of Strategic Change*, Oxford: Blackwell.

Hitt, M. and Ireland, D. (2000) 'The intersection of entrepreneurship and strategic management', in D. Sexton and H. Landstrom (eds) *Handbook of Entrepreneurship*, Oxford: Blackwell.

Hornsby, J., Naffziger, D., Kuratko, D. and Montagno, R. (1993) 'An interactive model of the corporate entrepreneurial process', *Entrepreneurship: Theory and Practice* 17, 2: 29–47.

Hussey, D. (1998) *How to Be Better At Managing Change*, London: Kogan Page.

Kanter, R.M. (1983) *The Change Masters: Corporate Entrepreneurs At Work,* London: Unwin Hyman.

Kanter, R.M. (2000) 'When a thousand flowers bloom: structural, collective and social conditions for innovation in organization', in R. Swedberg (ed.) *Entrepreneurship: The Social Science View*, Oxford: Oxford University Press.

Katz, R. (1994) 'Managing high performance R&D teams', *European Management Journal* 12, 3: 243–252.

Kondra, A.Z. and Hinings, C.R. (1998) 'Organizational diversity and change in institutional theory', *Organization Studies* 19, 5: 743–767.

Leigh, A. and Walters, M. (1998) *Effective Change: Twenty Ways to Make it Happen*, London: IPD.

Leonard-Barton, D. (1994) 'Core capabilities and core rigidities: a paradox in managing new product development', *Strategic Management Journal* 13: 111–125.

Levesque, P. (1999) *Breakaway Planning – 8 Big Questions to Guide Organizational Change*, London: Chartered Institute of Purchasing and Supply.

Lewin, K. (1947) 'Frontiers in group dynamics', *Human Relations* 1: 5–41.

Lippitt, R., Watson, J. and Westley, B. (1958) *The Dynamics of Planned Change*, New York: Harcourt, Brace & World.

Lumpkin, G.T. and Dess, G.G. (1996) 'Clarifying the entrepreneurial orientation construct and linking it to performance', *Academy of Management Review* 21, 1: 135–172.

Meyer, J.W. and Rowan, B. (1977) 'Institutionalized organizations: formal structures as myth and ceremony', *American Journal of Sociology* 83: 340–363.

Nelson, R. and Winter, S. (1982) *An Evolutionary Theory of Economic Change*, Boston, MA: Harvard University Press.

Orlikowski, W. (1992) 'The duality of technology: rethinking the concept of technology in organizations', *Organization Science* 3, 3: 398–427.

Pettigrew, A. (1992) 'On studying managerial elites', *Strategic Management Journal* 13 (Special Issue): 168–182.

Pfeffer, J. (1981) *Power in Organizations*, Marshfield, MA: Pitman.

Pinchot, G. (1985) *Intrapreneuring: Why You Don't Have to Leave the Corporation to Become an Entrepreneur,* New York: Harper & Row.

Powell, W.W. and DiMaggio, P.J. (1991) *The New Institutionalism in Organizational Analysis*, Chicago, IL: Chicago University Press.

Prior, A. and Kirby, M. (1998) 'The Society of Friends and Business Culture, 1700–1830', in D.J. Jeremy (ed.) *Religion, Business and Wealth in Modern Britain*, London: Routledge.

Pugh, D. and Hickson, D. (1976) *Organisational Structure and its Context: The Aston Programme I*, Farnborough: Saxon House.

Rogers, E.M. and Shoemaker, F.F. (1971) *Communication of Innovations: A Cross-cultural Approach* (2nd edn), New York: Free Press.

Rye, C. (1996) *Change Management Action Kit*, London: Kogan Page.

Schumpeter, J.A. (1934) *The Theory of Economic Development*, Cambridge, MA: Harvard University Press.

Scott, W.R. (1991) 'Unpacking institutional arguments', in W.W. Powell and P.J. DiMaggio (eds) *The New Institutionalism in Organization Theory*, Chicago, IL: Chicago University Press.

Scott, W.R. (1994) 'Institutions and organizations: towards a theoretical synthesis', in W.R. Scott and J.W. Meyer (eds) *Institutional Environments and Organizations*, Thousand Oaks, CA: Sage.

Scott, W.R. (1995) *Institutions and Organizations*, Thousand Oaks, CA: Sage.

Shane, S.A. (1994) 'Are champions different from non-champions?', *Journal of Business Venturing* 9, 5: 397–421.

Sharma, P. and Christian, J. (1999) 'Towards a reconciliation of the definitional issues in the field of corporate entrepreneurship', *Entrepreneurship: Theory and Practice* 23, 3: 11–28

Spender, J-C. (1996) 'Making knowledge the basis of a dynamic theory of the firm', *Strategic Management Journal* 17 (Special Issue): 45–62.

Stein, R.G. and Pinchot, G. (1998) 'Are you innovative?', *Association Management* 50, 2: 74–77.

Stevenson, H.H. and Jarillo, P.M. (1990) 'A paradigm of entrepreneurship: entrepreneurial management', *Strategic Management Journal* 11 (Special Issue): 17–27.

Stopford, J.M. and Baden-Fuller, C. (1994) 'Creating corporate entrepreneurship', *Strategic Management Journal* 15, 7: 521–536.

Tichy, N.M. (1975) 'How different types of change agents diagnose organizations', *Human Relations* 23, 5: 771–779.

Tolbert, P.S. and Zucker, L.G. (1983) 'Institutions as sources of change in the formal structure of organizations: the diffusion of civil reform', *Administrative Science Quarterly* 28: 22–39.

Welbourne, T.M. and Andrews, A.O. (1996) 'Predicting the performance of initial public offering: should human resource management be in the equation?', *Academy of Management Journal* 39, 4: 891–919.

Whittington, R. (1992) *What is Strategy and Does It Matter?*, London: Routledge.

Woodward, J. (1958) *Management and Technology*, London: HMSO.

Yin, R.K. (1994) *Case Study Research: Design and Methods* (2nd edn), Thousand Oaks, CA: Sage.

Zahra, S.A. (1996) 'Governance, ownership and corporate entrepreneurship: the moderating impact of industry technological opportunities', *Academy of Management Journal* 39, 6: 1713–1735.

Zahra, S.A. and Pearce, J.A. (1994) 'Corporate entrepreneurship in smaller firms: the role of environment, strategy and organization', *Entrepreneurship, Innovation and Change* 3, 1: 31–44.

Zucker, L. (1983) 'Organizations as institutions', in S.B. Bacharach (ed.) *Research in the Sociology of Organisations*, Greenwich, CT: JAI Press.

8 Experimentation as a boundary practice in exploring technological development processes in chemistry

Sari Yli-Kauhaluoma

Introduction

Technological change in chemistry evolves through the development of new chemical products and processes. In this technological development process, experimentation is one of the most important practices (cf. e.g. Holmes and Levere 2000). This means that experimentation is crucial for creating new technological knowledge about chemical materials and chemical reactions. Importantly, it is not a single, stable organizational activity. Instead, as previous research has argued (D'Adderio 2001), it consists of a cluster of activities which are closely linked together. One of the most recent studies dealing with various kinds of experimentation is by Thomke (1998: 744). He has identified that experimentation consists of four main activities which together comprise an iterative learning cycle. These activities include designing an experiment, building an apparatus to conduct the experiment, running the experiment and analysing the results.

Experimental activities are important because they constitute organizational action and therefore help us to understand how technological change unfolds. So far, such experimental activities, their premises and consequences, have attracted relatively little attention (D'Adderio 2001). This does not mean that experimentation has not been studied. On the contrary, earlier literature has addressed various aspects of the subject. These include different modes (Thomke 1998) and speeds (Thomke *et al.* 1998), uncertainty (Fleming 2001) and the amounts of resources needed (Dahan and Mendelson 2001), and the various strategies involved (Loch *et al.* 2001). Altogether, however, the focus of previous literature on experimentation has largely been on the management of problem-solving related to technological change (Rosenberg 1982). In other words, a wide range of literature has been interested mainly in how to manage a technological development process as efficiently as possible. In recent years, however, a body of literature has been emerging with a significantly different perspective to the study of technological change. A practice-based theory (cf. e.g. Pentland 1992; Orlikowski 1996, 2000; Orr 1996; Tyre and von Hippel 1997) for the study of the varying dimensions of this organizational problem-solving activity has been developed. In other words, this practice lens (Orlikowski 2000) focuses on the study of organizational action in the use of technological artefacts. This means that the use of technology 'is not a choice among a closed set of predefined possibilities' (Orlikowski 2000: 409). Instead, it is enacted in context. This means that organizational action attached to use of a technology, for instance, experimentation, is the focus of studying the development of a technology. Thus, this practice perspective seems useful since it allows us to explore various organizational activities in the process of technological change (cf. Barley 1986; McLoughlin and Dawson, Chapter 2, this volume).

This chapter examines technological experimentation in the context of chemistry. The focus is on the organizational action involved in the technological change process for a new chemical technology. Based on an empirical case study, I use Thomke's (1998) above-mentioned elements of experimentation as the starting point of the analysis. The results suggest that each of these elements consists of several organizational activities that link different, formally separated organizational actors with each other. This means that experimentation is not merely an organizational problem-solving activity. It is also an activity that helps to bridge boundaries between different organizations during the development of a technology.

The purpose of this chapter is to study the organizational action involved in a technological change process. The focus is on experimentation in the context of chemistry. The aim is to show that experimentation is not only an organizational problem-solving activity, as conventionally regarded, but it also plays an important role in bringing formally separated organizations together. Therefore, the remainder of the chapter consists of the following. First, I review briefly the previous literature on experimentation and focus on the perspectives offered by the practice-based perspective. Second, I present the research setting and methodology, after which I move on to analyse the various dimensions of experimentation in this case. Finally, I summarize and discuss the research findings.

Perspectives on experimentation

There is a wide body of research (cf. e.g. Thomke *et al.* 1998; Dahan and Mendelson 2001; Loch *et al.* 2001) that approaches experimentation as a technical problem-solving activity. Consequently, the question of how to arrange successful experimentation has received most attention in the previous literature. One of the liveliest debates in this domain has concerned the different modes of experimentation and some of the consequences of choosing a particular mode. The two main modes that have been identified in the literature are serial and parallel experimentation (Thomke 1998). The main difference between these two modes lies in the role of learning from previous tests. Tests that incorporate learning from other experiments are called serial experimentation strategies. In contrast, parallel experiments are tests that follow a pre-established plan without modifications derived from the findings of other tests (Thomke *et al.* 1998: 318).

The choice between the serial and parallel modes is considered relevant first and foremost because it affects both the speed and the cost of experimentation. This finding has led to studies that aim to model optimal testing strategies (e.g. Loch *et al.* 2001). Other attempts in the same spirit have pointed to the question of the resources needed for experimentation (Dahan and Mendelson 2001). Finally, the aspect of risk-taking in experimentation has also been studied. Fleming (2001), for instance, argued that although a test organization that contains components and combinations that are novel and uncertain to inventors often leads to failures, it may also increase the potential for technological breakthroughs.

In recent years, however, another research approach to the study of technological change that can be used to extend previous research on experimentation has emerged. This approach is called the practice-based perspective and emphasizes various dimensions of organizational problem-solving activity (cf. e.g. Pentland 1992; Orlikowski 1996, 2000; Orr 1996; Tyre and von Hippel 1997). In contrast to mainstream research on experimentation, the focus here is on organizational action. Attention is paid to the activities people carry out with the technology and how they use it in a specific context. It is through these activities that technological change unfolds. Thus it is not so much a question of arranging a technological problem-solving activity as exploring those organizational problem-solving activities through which technologies change. An advantage of this approach is that it is sensitive to contextual matters of technological change. In addition, the practice-based perspective allows us to capture some of the dynamic elements in the process of developing technology. This means that instead of seeing technological problem-solving as a single activity, it sees it instead as consisting of multiple activities. Further, it is important to note that these activities are not necessarily stable. Instead, as recent literature has pointed out (Knorr Cetina 2001: 175), in particular the activities needed to confront 'non-routine problems' are in constant flux. Previous research has identified several dimensions of experimentation as a set of organizational activity. According to Thomke (1998), experimentation consists of four main activities (i.e. designing an experiment, building an apparatus to conduct the experiment, running the experiment and analysing the results). Together these activities create an iterative learning cycle. Pisano (1994) has pointed to another perspective important in exploring experimentation as organizational action. His argument is that the environment of experimentation plays a crucial role by varying the dimensions of organizational problem-solving activities. Pisano's work is especially relevant here because he has also examined chemistry and identified three main environments where experimentation concerning technological change takes place. In chemistry, these environments are laboratory, pilot plant and full-scale commercial factory levels. According to Pisano (1996: 1105), some of the main dimensions that affect organizational activity in each of these environments are the type of equipment used, the number of steps, the criteria for design, the output and the operators.

Another interesting study tackling the role of the physical environment in various organizational dimensions of problem-solving activities was conducted by Tyre and von Hippel (1997). Their argument is that when people move back and forth between different physical environments, they confront different sorts of clues and gather important information embedded in these environments. These clues and information are needed to solve the technical problems of the various phases of experimentation. An opposite perspective to experimentation is offered by Pentland (1992), whose study on software support hot lines demonstrates how technical specialists create solutions to complex technical problems. These solutions are then tested online in practice by the practitioners.

Altogether, the studies of Pisano (1994, 1996), Tyre and von Hippel (1997) and Pentland (1992) were presented above mainly because they all point to an important but largely omitted aspect in technological experimentation. That is, they all show how organizational action involved in the change of a particular technology takes place simultaneously in multiple environments. In the same spirit, experimentation as a problem-solving activity can take place across organizational units or even across organizational boundaries. When this is the case, it is the bridging mechanisms between different loci that become crucial for understanding technological change. Previous literature has recognized boundary spanners (Tushman 1977), boundary objects (Leigh Star and Griesemer 1989), and recently, boundary practice (Wenger 1998) as possible bridging mechanisms between separate organizations. The idea of boundary practice is particularly appealing here because it allows us to think about experimentation as a bridging mechanism between organizations in a technological change process. This suggests that experimentation, as organizational action, has a double meaning in the development of technology. In addition to viewing experimentation conventionally as a problem-solving activity, we can also see it as a bridging mechanism that creates necessary linkages between various organizational actors in the process of the development of a technology. Through this case study material, I aim to elaborate this double meaning of experimentation.

Research setting and process

The empirical material presented in this chapter is based on a qualitative, retrospective single case study on the development of a novel chemical technology, a new type of recyclable catalyst. Catalysts are materials that alter the rate of attainment of chemical equilibrium without themselves being consumed in the process. They are one of the core technologies for the manufacture of chemicals and materials and are widely used in many different industries, from food processing to pharmaceuticals (Wittcoff and

Reuben 1996; Adams 1999). In this specific case, the catalyst was developed for the production of novel ingredients used in functional food. Functional food products can briefly be defined as food products which contain ingredients with beneficial health impacts that go beyond normal nutrition. The key invention in the catalyst was to use a new type of material – fibre – instead of carbon as the essential component of the catalyst. This new type of material made recycling of the catalyst possible. Thanks to the fibre, the catalyst was also more effective.

The focus of the empirical analysis is on the various dimensions of experimentation during the technological change process of the catalyst. The transition phase, in which the catalyst moves from a research laboratory to industrial settings, provides the context for the empirical analysis. This means that we follow a process in which the catalyst first existed only in a research laboratory where it was used for small-scale scientific experiments. Consequently, many different types of experimentation with the catalyst were required for it to become applicable in large-scale production processes. Through detailed examination of this experimentation, I aim to illustrate and elaborate further the conceptual framework.

At the level of individuals, there were eleven chemists, chemical engineers and chemical technicians in five separate organizations involved in the technological development process for the catalyst (see Table 8.1). These organizations consisted of two chemical companies (Catco and Chemco), the paper and pulp mill of a major forest industry corporation (Woodco), a food-processing firm (Foodco) and a department at a technical university (Tech). Of the chemical companies, Catco was a start-up firm that the original inventors of the idea set up around the catalyst. Chemco was an established firm. Altogether, the transition phase from laboratory into industry and thus the experimentation period took about four years, from the beginning of 1992 until the summer of 1995. This period also provides the time frame for the empirical analysis. All the names of the organizations involved are pseudonyms.

The main empirical material of the case study consists of twelve interviews made between June 1999 and November 2000. The empirical material also includes documents concerning chemical experiments on the catalyst, scientific

Table 8.1 Organizations and individuals involved in the development of the catalyst

Pseudonym of the company	*Industry*	*Type of organization*	*Number of key persons involved*
Catco	Chemistry	Start-up company	3
Chemco	Chemistry	Large established firm	2
Woodco	Paper and pulp	A mill belonging to a major corporation	2
Foodco	Food processing	Large established firm	2
Tech	Academe	University department	2

articles published on the properties and mechanisms of the catalyst, and industrial seminar material concerning the experimentation in a transition phase from laboratory to industry. Out of eleven key chemistry professionals who were concretely involved in experimentation with the catalyst, ten people representing five different organizations were interviewed. These key interviewees were identified in the course of the interview process. In four interviews, there was more than one interviewee. As a result, I met most of the interviewees at least twice. In addition to successive interviews, the discussion continued either by phone or through email. In the interviews, I followed the suggestion of Lynn *et al.* (1996) and concentrated on the commercialized technology, in this case the catalyst. This means that all interviews took the form of discussions on how this specific catalyst was developed. The interviewees explained how their practices varied as work on the catalyst proceeded.

Experimentation cycle of the catalyst

In the case in focus, a group of university scientists invented a new type of catalyst. The main novelty of the catalyst was the use of fibre as its key material. Soon after the group had filed a patent application for the new catalyst, it set up a company and decided to pursue commercialization of the invention. Commercialization required numerous experiments in different physical settings and locations to make the transition from laboratory to industrial setting possible. Here, the focus is on an analysis on the variety of experimentation carried out by different organizational actors during a technological change process that ended in large-scale production of not only the catalyst, but also a novel functional food ingredient where the catalyst was used in a chemical reaction. This suggests that experimentation should be seen not only as a problem-solving activity but also as a linking mechanism between actors representing separate organizations. Table 8.2 summarizes the findings from the empirical material.

Designing an experiment

Determining the object and purpose of testing is one of the first prerequisites for designing an experiment. In the case of the catalyst, this meant that the first problem faced by Catco, the start-up company, in the commercialization process was to find potential applications for the catalyst. Catco approached this problem by creating opportunities for other organizations to design and run tests of its new technological invention. In practice, Catco delivered samples of its catalyst to numerous organizations, which were then able to design tests for their own technological problems. At the same time, however, Catco itself also explored potential industrial processes where its invention could be put to use. It contacted firms and offered to arrange tests of the catalyst concerning chemical reactions that it considered interesting and critical for the contacted firms.

Table 8.2 Experimentation as a problem-solving and bridging activity in the case of the catalyst

Stage in the experimentation cycle	*Problem-solving activities*	*Bridging activities*
Designing tests	Searching for potential applications of the catalyst	Exploring potentially interesting industrial processes for separate organizations
	Studying the scientific premises of the catalyst	Delivering samples of the catalyst to different organizations
	Studying the scientific premises of the chemical reaction	Carrying out joint research projects
Building apparatus	Searching for proper machines, facilities and knowledgeable operators	Using and modifying existing machines
		Renting facilities
	Building a production line for the catalyst	Donating and transferring pieces of equipment
Running experiments	Fine-tuning of the catalyst for a particular chemical reaction	Disseminating knowledge concerning the catalyst between separate organizations
	Studying the effects of scale	Arranging tests in various groupings of organizations
		Arranging tests in different physical settings
Analysing results	Making decisions concerning the next step	Creating and preventing opportunities for further linkages

The contacts and co-operation needed to design an experiment and then test the catalyst accordingly in chemical reactions facilitated linkages between Catco and other involved organizations. It also led to the emergence of the technological development process of the catalyst, which was to be applied in the production of a functional food product. In practice, this process started from two separate and successful test designs and experiments. First, Catco proposed a test design and experiment to Foodco, a food-processing firm. The experiment resembled a chemical reaction needed in Foodco's functional food production. This experiment turned out to be successful. It also provided an opportunity to proceed to experiments where the catalyst was used in the production of a functional food product. Second, Catco and Tech, the technical university, conducted a joint research project on the initial stage in the development of the catalyst. More specifically, Catco had delivered some samples of the catalyst to Tech for experimentation and study in a Master's thesis. It was also important, however, that Tech was simultaneously carrying out another research project with a forest industry firm, Woodco, where it learned from the latter about its difficulties in tests concerning production of functional food ingredients. This led to a proposal to design and conduct experiments with Catco's catalyst in a chemical reaction critical for functional food production. In any case, however, the details mentioned indicate that designing and conducting experiments, in this case with the catalyst in different chemical reactions, created linkages between various organizational actors. It also triggered a technological development process that later led to use of the catalyst in the specific chemical reaction needed for production of functional food ingredients.

In the course of developing the catalyst, experiments were needed to solve problems in two main areas. First, there was a need for experiments to test the properties and mechanisms of the catalyst. This was mainly because the key material in the catalyst, fibre, had not been used before. The experiments were also needed to test the catalyst in a practical situation; that is, in a chemical reaction for the production of functional food ingredients. These two different types of experiments, namely the properties and mechanisms of the catalyst and the chemical reaction where the catalyst was used, were not two separate phases. In fact, they were closely linked. This means that problems in testing the catalyst in a specific chemical reaction implied reacting not only in the form of test designs of the catalyst in the chemical reaction, but also in the form of test designs of the properties and mechanisms of the catalyst. Put differently, various organizational actors were linked to each other through different types of experimentation. For example, in the process of developing the catalyst, there appeared to be some severe problems with the key material, fibre, in a chemical reaction. In some tests, the fibre simply disintegrated. This was not supposed to happen in any circumstances. It was at Tech where this problem was first noticed. As a result, the tests were redesigned both at Tech and also at a large chemical company, Chemco, which had developed and delivered the special type of fibre material to Catco. The new test designs at Tech concerned the chemical reaction where the catalyst was used, while Chemco focused on the particular properties and mechanisms of the catalyst, fibre. These details indicate that the experimentation by the organizations involved did not comprise separate actions, but closely related ones instead. Moreover, knowledge from previous experiments was used by the actors involved.

Building an apparatus to conduct an experiment

In chemistry, three types of equipment are needed typically to conduct experiments concerning commercialization of chemical products and processes. The key difference between the types of equipment lies in their scale. Some of the pieces of equipment are laboratory-scale machines. The pilot-scale equipment used in experiments is already much larger than that used in laboratories.

Eventually, full-scale industrial production scale equipment is required for the final stages of the testing process. Although chemical products and processes all go through these steps from laboratory via pilot plant to full-scale industrial production, this does not mean that different types and sizes of apparatuses had to be built each time to carry out tests in all of these settings. This is because the equipment is not only expensive, but also requires proper facilities and knowledgeable operators who are able to run different types of machines. In fact, different scales of machines are often located in different kinds of organizations. As experimentation proceeds, linkages between various formally separated organizational actors emerge and machines are either used as such or modified to suit the requirements of the experiments. In the case of testing the catalyst in focus, laboratory-scale equipment was located in every organization. Nonetheless, most of these tests were run at either Catco or Tech. Pilot-scale apparatus was located only in the large-scale organizations (i.e. at Woodco and Foodco). None of the organizations involved in the development process possessed a full-scale industrial production line for a functional food ingredient to which the catalyst could be applied. Therefore, such facilities were rented. Hence, it seems justified to argue that different types and scales of experiments acted as bridging mechanisms between various organizations involved in the technological development process of the catalyst. The locus of the equipment has some important consequences for experimentation. Knowledge of the design of experiments using particular equipment and experience in operating certain machines is not easily or quickly acquired. This is especially the case in organizations that do not possess those particular machines. Technological development projects often have tight schedules. Therefore, contacts to organizations where such knowledge and pieces of equipment already exist are invaluable. This was also the case concerning the catalyst. As the fibre disintegrated in the experiments, Catco was able to quickly contact Chemco, which designed and ran the necessary experiments for Catco.

Although most of the machines and equipment needed in the experiments with the catalyst already existed at the premises of some of the organizations involved, there was also a need to build new apparatus. The most important of these were actually the machines that comprised the production line for the catalyst. A production line was built because large amounts of the catalyst were first needed in the larger scale experiments and later in the eventual production process. It was possible to produce the necessary amount of catalyst needed in small-scale tests by hand in laboratory conditions. However, larger scale experiments starting from the pilot plant phase required large amounts of catalyst. It was not possible to produce such amounts by hand. Instead, different types of machines, equipment and apparatus were needed. Catco designed and built this production line, but even in this phase its connections with other organizations played a crucial role. Catco received different types of assistance and guidance concerning the construction of a production line from Chemco. Chemco donated and transferred some of its old machines to Catco and helped it to use this equipment. Thus, in the construction of the production line for the catalyst, previous experimentation with special machines played a crucial role and linked two organizations: Catco and Chemco.

Running an experiment

The focus of problem-solving in experimentation is not stable; it evolves during the technological change process. In the case of the catalyst, the first experiments concerned industrial processes to which this new type of catalyst could be applied. As soon as one chemical reaction was identified, the focus of experimentation shifted to fine-tuning of the basic applicability of the catalyst in this chemical reaction. Both the properties and mechanisms of the catalyst and its applicability in a specific chemical reaction played a crucial role in testing. Moreover, the focus of experimentation shifted to identifying and solving any problems concerning the effects of scale only after the key problems in these two areas had been solved. In practice, this meant that the experimentation expanded into different types of physical setting that were mainly distinguished from previous test environments by the scale of the experiments. After the main problems in these areas had been solved, testing of the catalyst in a specific chemical reaction expanded into the final application setting and to possible problems there.

Understanding this changing focus of experimentation is important because it is involved with the shifting organizational locus of the experimentation. In the case of the catalyst, it was not only a specific type of experiment that was modified as a result of learning from previous steps. In addition, it was the type and physical setting of the test that changed in the technological development process. This meant that during the evolving technological testing process there were important changes in the composition of the organizations responsible for carrying out the tests. Various tests concerning the catalyst as a potential application in industrial processes were carried out in several organizations. Experiments at Catco and Tech were therefore critical, as a decision was made to continue testing of the catalyst in a specific chemical reaction used in a functional food production process. Tests run particularly at Chemco, Tech and Catco played a crucial role before the experimentation proceeded to tests concerning the effect of scale, which was tested at both Woodco and Foodco. Thus different types of testing acted as a bridge between different organizations in a technological change process.

Interestingly, running tests also took place in various groupings of the organizations participating in the process. Collaboration between separate organizations involved in experimentation was thus closer in some cases than in others. In practice, this led to a situation where similar types of tests were carried out simultaneously in different groups of organizations. For instance, there was a close linkage between Catco and Foodco in organizing, running and analysing both laboratory and pilot-plant type tests concerning the applicability of the catalyst to a specific chemical reaction. A similar type of linkage existed between Tech and Woodco. One reason for this may have been the geographical location. For example, the facilities of Catco and Foodco were located only a few kilometres from each other. The same geographical proximity characterized Tech and Woodco, while the distance between these and the other two organizations was several hundred kilometres.

Finally, running tests may be seen as an important way to disseminate knowledge to new organizations and thus to create opportunities for new interorganizational linkages to emerge. In the case of the catalyst, Catco and Tech had a joint research project concerning the initial stage in the development of the catalyst. This was before the start of the development process concerning its applicability for functional food production. While Tech was running tests concerning the initial stage of the catalyst, Tech gained valuable knowledge about its scientific premises and functioning. This knowledge turned out to be important, as Tech learned about Woodco's problems in the production of functional food ingredients. Importantly, this knowledge was crucial, not only because it offered a solution to Woodco's problems, but also because new knowledge enabled Tech to join the technological development process of a catalyst for a functional food production process.

Analysing the results of an experiment

The decision concerning the next step in the experimentation is based on analysis of the results from previous tests. Basically, three different types of decision can be made. First, analysis of the results may show that some of the problems originally confronted still exist. Therefore, more similar types of testing are needed, and the test designs need modification. The second decision may be that although the problem was solved, some other problems exist or even new ones emerged based on the solution to the previous problem. This means that different types of testing, perhaps also in different test settings, are required. The third possible decision is that no further testing is carried out. This may be because the solution is viable enough or simply because there are no possibilities to continue testing, for one reason or another.

All three different types of decision refer to the problem-solving aspect of experimentation. However, based on knowledge from this study on the catalyst, the analysis of the results may also lead to different kinds of consequences. Namely, it may either create or prevent opportunities for further linkages between different organizations. One example of such an emerging linkage was the analysis by Foodco of the test results carried out by Catco. These test results concerned a chemical reaction which was similar to a type of reaction that was particularly interesting to Foodco. Foodco analysed the tests made at Catco and realized that there was potential for developing this new type of catalyst or a technology for use in functional food production. Thus the results of an experiment led to the continuation and enhancement of a linkage between Catco and Foodco.

As mentioned above, however, analysing the results of an experiment may also prevent linkages between different organizations from developing further. In the case in focus, as the technological development of the catalyst had proceeded to the final stage, industrial scale facilities were needed for the production of functional food ingredients applying the catalyst. Therefore, Foodco and Catco conducted several negotiations with different potentially suitable organizations and proposed test runs concerning the catalyst in a specific chemical reaction. As Foodco and Catco then analysed the results of these test experiments, they noticed in some cases that the results were simply not good enough. These results showed that the testing organization had not been able to carry out the test properly and therefore negotiations with these organizations were not continued. Thus, analysing the results of an experiment had prevented further development of linkages between Foodco, Catco and some of the testing organizations.

Finally, concerning the experimentation cycle, this case suggests that analysing the results of an experiment is a specific skill requiring a particular kind of knowledge and expertise. This knowledge is often tied to using a particular type of equipment required for analysing the tests. Therefore, linkages to organizations with the equipment and knowledge needed to analyse the results become crucial. In the case of the catalyst, the analyses of the tests run at Catco, for instance, were often carried out at Foodco, since the latter had the expertise needed for different types of analysis.

Conclusions

Previous literature has acknowledged the importance of experimentation in technological development processes in various contexts such as chemistry (cf. e.g. Holmes and Levere 2000). So far, however, the existing research has focused mainly on efficient and successful ways of managing and arranging experimentation. It is only recently that a different approach has emerged to challenge and extend this view. A practice-based perspective (cf. e.g. Pentland 1992; Orlikowski 1996, 2000; Orr 1996; Tyre and von Hippel 1997) draws our attention to various dimensions of organizational action. Thus, the focus of the practice-based perspective is on those activities that people do with the technology, and how they use a technology in a specific context. One of the advantages of this perspective is that it allows us to capture at least some of the dynamic elements in technological change processes.

Recent research has shown that experimentation consists of several organizational activities (Thomke 1998). It has been thought that the main aim of these activities is to solve the more or less concrete problems in technological development processes. The previous research has also pointed out that the specificities of the context play an important part in organizational problem-solving activities (cf. e.g. Tyre and von Hippel 1997). This is particularly the case in chemistry, where different scales of experiment require various physical settings for experimentation (Pisano 1994, 1996).

Because of these multiple physical settings, the bridging mechanisms between the various environments become crucial in the unfolding of technological change. Boundary practice (Wenger 1998), which refers to activities that link separate organizations, has been identified as one of these bridging mechanisms. For this study, this idea is appealing since it allows us to think of experimentation not only in the traditional way, as a problem-solving activity, but also as a bridging mechanism between separate organ-izations. The empirical evidence provided in this chapter supports this perspective. Hence experimentation has a double meaning in the development of technology. In the case presented in this chapter, the analysis of experimentation using several analytical dimensions showed that in addition to being a crucial problem-solving activity, experimentation is also an important linking mechanism between the actors involved in a technological development process. Apparently, more research is needed to elaborate further the mechanisms through which the linking effect between different organizations occurs and the effects of this linking in different settings. In particular, results from this study suggest that further work should be conducted in two directions. First, while we noted that experimentation creates opportunities for linkages between different organizations, it may also prevent emerging linkages from being materialized. On the one hand, for instance, organizing different types of experimentation, which were needed during a technological change process, was an important element in the process of linking organizations. It even helped different organizations to become involved in development of the catalyst. On the other hand, analysing the results of experiments may point not only to problems or solutions in the development of a technology, but also indicate whether the testing organization is capable of carrying out experiments in the first place. Second, the closeness of collaboration between different organizations involved in the experimentation varies. This means that although several organizations were basically developing the same technology, the connections between some organizations may be much closer than those between certain others. Finally, this study confirms that experimentation remains one of the key problem-solving activities in chemistry. There are various kinds of test needed to develop a particular technology. In addition, testing has to proceed through various physical settings. While this chapter has focused on chemistry, studies of other industries will provide further insight into the realm of social bridging between organizations triggered by experimentation.

Acknowledgements

I would like to thank Juha Laurila, Hannu Hänninen, Juha Laaksonen, and David Preece for their helpful comments on earlier drafts of this chapter. This chapter draws on my doctoral research, which was supported in part by the Helsinki School of Economics and its foundations, and the Academy of Finland. Additional support from the Foundation for Economic Education is also gratefully acknowledged.

References

Adams, C. (1999) 'Catalysing business', *Chemistry & Industry* 19: 740–742.

Barley, S.R. (1986) 'Technology as an occasion for structuring evidence from observations of CT scanners and the social order of radiology departments', *Administrative Science Quarterly* 31: 78–108.

D'Adderio, L. (2001) 'Crafting the virtual prototype: how firms integrate knowledge and capabilities across organizational boundaries', *Research Policy* 30: 1409–1424.

Dahan, E. and Mendelson, H. (2001) 'An extreme-value model of concept testing', *Management Science* 47, 1: 102–116.

Fleming, L. (2001) 'Recombinant uncertainty in technological search', *Management Science* 47, 1: 117–132.

Holmes, F.L. and Levere, T.H. (eds) (2000) *Instruments and Experimentation in the History of Chemistry*, Cambridge, MA: MIT Press.

Knorr Cetina, K. (2001) 'Objectual practice', in T.R. Schatzki, K. Knorr Cetina and E. von Savigny (eds) *The Practice Turn in Contemporary Theory,* London: Routledge.

Leigh Star, S. and Griesemer, J.R. (1989) 'Institutional ecology, "translations" and boundary objects: amateurs and professionals in Berkeley's Museum of Vertebrate Zoology, 1907–39', *Social Studies of Science* 19: 387–420.

Loch, C.H., Terwiesch, C. and Thomke, S. (2001) 'Parallel and sequential testing of design alternatives', *Management Science* 45, 5: 663–678.

Lynn, L.H., Mohan Reddy, N. and Aram, J.D. (1996) 'Linking technology and institutions: the innovation community framework', *Research Policy* 25: 91–106.

Orlikowski, W.J. (1996) 'Improvising organizational transformation over time: a situated change perspective', *Information Systems Research* 7, 1: 63–92.

Orlikowski, W.J. (2000) 'Using technology and constituting structures: a practice lens for studying technology in organizations', *Organization Science* 11, 4: 404–428.

Orr, J. (1996) *Talking about Machines*, Ithaca, NY: Cornell University Press.

Pentland, B.T. (1992) 'Organizing moves in software support hot lines', *Administrative Science Quarterly* 37: 527–548.

Pisano, G.P. (1994) 'Knowledge, integration, and the locus of learning: an empirical analysis of process development', *Strategic Management Journal* 15: 85–100.

Pisano, G.P. (1996) 'Learning-before-doing in the development of new process technology', *Research Policy* 25: 1097–1119.

Rosenberg, N. (1982) *Inside the Black Box*, Cambridge, MA: Cambridge University Press.

Thomke, S., von Hippel, E. and Franke, R. (1998) 'Modes of experimentation: an innovation process – and competitive – variable', *Research Policy* 27: 315–332.

Thomke, S.H. (1998) 'Managing experimentation in the design of new products', *Management Science* 44, 6: 743–762.

Tushman, M.L. (1977) 'Special boundary roles in the innovation process', *Administrative Science Quarterly* 22: 587–605.

Tyre, M.J. and von Hippel, E. (1997) 'The situated nature of adaptive learning in organizations', *Organization Science* 8, 1: 71–83.

Wenger, E. (1998) *Communities of Practice*, New York: Cambridge University Press.

Wittcoff, H.A. and Reuben, B.G. (1996) *Industrial Organic Chemicals*, New York: Wiley.

9 Regional technology systems or global networks?

The sources of innovation in opto-electronics in Wales and Thuringia

Chris Hendry, James Brown, Hans-Dieter Ganter and Susanne Hilland

Introduction

The central question in this chapter is the extent to which national (and by extension regional) systems of innovation are reduced in influence in a technological domain that is highly scientific and international in its very nature.

Recent work on high-technology regional clusters and industrial districts has tended to emphasize the need for firms to be open to global networking (Cooke *et al.* 1995) and for analysis of regional clusters to take account of the way these are embedded in international networks (Amin and Thrift 1992; Garnsey and Cannon-Brookes 1993; Hahn and Gaiser 1994; Keeble 1994; Ganter 1997). Thus Scott (1993: 26–27) argues that 'much of the contemporary world economy can be seen as a mosaic of regional agglomerations (marked by localized transactional networks) embedded in far-flung systems of national and international transacting'. Similarly, high-performing city regions are 'increasingly embedded in global networks of production and exchange' (Staber 1996). On the other hand, national systems of innovation acquire localized characteristics, and regional systems of innovation also merit consideration (Casper and Vitols 1997; Cooke *et al.* 1997).

A recent study of opto-electronics firms in six clusters in the UK, Germany and USA confirmed this pattern of global networking reaching out from the region (Hendry *et al.* 1999, 2000a). However, it also revealed significant differences in the development of these six regions as industrial districts – that is, in the extent of local inter-firm networking, the extent of local research collaborations (involving firms, universities and other research centres), the provision of a regional support infrastructure, and culturally embedded social networks. A subsequent survey, covering all opto-electronics firms in the UK, showed these globalizing tendencies to be characteristic of UK firms as a whole (Hendry and Brown 1999), whether located in clusters or dispersed across the country.

Nevertheless, some regions do have a stronger inward focus in terms of trading patterns and collaboration between companies and with local research centres. Of the six clusters studied (East Anglia, Wales, Massachusetts, Arizona, Munich and Thuringia), the district around Jena (Thuringia) can lay claim to being the most developed as a classic industrial district (Becattini 1990). Wales, on the other hand, is in many respects the antithesis in its pattern of evolution. Opto-electronics firms in Wales look outside the region for their customers, suppliers and collaborators, while Thuringian firms are focused strongly on local suppliers and collaborators.

These differences stem from the origins and character of these firms, and the recent history of the two regions. Thus Welsh firms are older, foreign-owned (to a greater extent), and the regional research infrastructure is less dense (Table 9.1). The age of firms relates to their involvement in markets that were among the first to develop (such as defence), and this, along with the opportunity to grow over a period of time, has encouraged a national and international orientation. At the same time, Wales pursued a policy of encouraging inward investment to compensate for the decline in its traditional industries (such as coal and steel) and this brought in multinational companies from overseas, as well as UK firms relocating to take advantage of grants and other incentives. In the process, they brought with them trading relationships and networks, and a mentality that transcended the locality.

In contrast, many opto-electronics firms in Thuringia are local spin-offs from Carl Zeiss Jena, or are new start-ups by former employees from Carl Zeiss and the University of Jena, following the breakup of the Carl Zeiss Kombinat post-reunification. Thus, although Carl Zeiss' history in the former GDR goes back to 1846, and it developed opto-electronics capability in lasers in the 1960s, the present population of firms is young, especially from the point of view that many have had to radically reorientate themselves to new markets. However, these firms and the public research infrastructure do retain fairly strong personal ties from their previous incarnation, and these have been reasserting themselves following the traumas of reorganization and redundancy. Finally, Thuringia, unlike Wales, has a very dense research infrastructure, and the Land government encourages firms to make use of this through funding for collaborative research projects between firms and the universities.

Notwithstanding these differences, both Wales and Thuringia have governmental agencies that have been pursuing active policies for regional development against a background of economic decline. The Welsh Development Agency (WDA) has hitherto been a relatively unique institution in the UK in this respect (Scotland also having a regional development agency), although, of course, the Länder in Germany are used to having this role. However, the policies of Wales and Thuringia are rather different.

However, whereas Thuringia has been focusing primarily on technology transfer and enterprise development, the WDA has given particular attention to international marketing and global alliances for innovation. These differences may be related to the specific circumstances of the two regions – in Wales, to the early emphasis on inward investment; in Thuringia, to the

Table 9.1 Demographic and institutional comparison between Thuringia and Wales

	Thuringia		*Wales*	
Area (km^2)	16,171		20,779	
Population	2,462,836 (1998 figure)		2,926,900 (1997 figure)	
Major towns	Erfurt – 204,000 Gera – 118,000 Jena – 99,000		Nearly 75% of population lives within 60 miles of Cardiff	
Employment	972,100 (1998 figure)		987,363 (1997 figure)	
	Breaks down as %		Breaks down as %	
	Government	17.4	Primary	3.7
	Manufacturing	19.7	Manufacturing	21.7
	Construction	5.8	Services	69.9
	Commerce	11.9	Other	4.7
	Services	18.8	But GDP breaks down as %	
	Other	16.4	Primary	3.3
			Manufacturing	29.3
			Energy and construction	8.4
			Private services	36.6
			Public services	22.4
Education and Research Institutions with examples of key centres in brackets	Heavy concentration in Erfurt and Jena		Mainly in South Wales but more distributed than Thuringia	
College or university	5 (FSU Jena)		15 (Cardiff, Swansea, Aberystwyth, Bangor)	
Non-academic research facility	9 (Fraunhofer, Max-Planck)		35 'centres of expertise' all but one or two in university departments	
Polytechnic	3 (Fachhochschule Jena)		28 institutions (11 FE colleges, 9 tertiary colleges)	
Business-oriented research institution	20 (CIS, MAZet)			
Technology and initiative centre	10 (TIP Jena)		14 science park or incubator centres	
Transfer centre	10 (THATI)		16 industrial liaison offices mainly in universities	
Average age (years) of opto-electronics case companies	8		17	

Sources:
Thuringian websites
Http://www.thueringen.de/index.html
Http://www.stift-thueringen.de/en/index.htm
Wales websites:
http://www.invest-in-wales.com/
http://www.wda.co.uk/

need to develop viable new enterprises from the fall-out from Carl Zeiss Jena and the public universities.

The contrasting experiences and policy emphasis of these two regions provide an opportunity therefore to compare the impact of national and local policies towards a particular sector, while taking account of its distinctive pattern of evolution locally (its 'path-dependent' character (Dosi 1982)). This leads to the question as to whether a regional policy based on linking firms into the global economy (WDA) is more viable, appropriate and sustainable than the development of a regional technology system (Thuringia), and what lessons for innovation may be derived from the experiences of firms under these two regimes. The research undertaken in 1999 to 2000 with the support of the Anglo-German Foundation for the Study of Industrial Society[1] attempted to answer this question. After briefly describing a model of innovation we outline the development of the opto-electronics industry with policy initiatives at the national and regional levels in the UK and Germany. We then set this against some evidence of how firms in the industry actually innovate. In this way we progressively refine the picture at an increasing level of detail, of what drives innovation in the two regions and how companies' innovation needs (and therefore what is relevant policy support) change over time.

Innovation in the modern world

Freeman (1974) describes innovation as a process, which includes 'the technical, design, manufacturing, management and commercial activities involved in the marketing of a new (or improved) product or the first use of a new (or improved) manufacturing process or equipment'. As this definition suggests, innovation does not only mean technological advance. For a new technology to be accepted by the market, it often needs to be accompanied by changes in R&D, production, commerce and marketing.

In defining a model of the innovation process, Rothwell (1992) moves beyond simple linear ideas of technology-push and market need-pull to provide a more general picture of the process of the coupling between science, technology and the marketplace. His 'interactive model' is depicted as 'a logically sequential, though not necessarily continuous process that can be divided into a series of functionally distinct but interacting and interdependent stages'. This is essentially a model of a networked process and Rothwell (1992) goes on, in more advanced versions, to introduce concurrent activities and simulated projections to arrive eventually at a systems integration and networking model. The essential point is that not only is technology changing rapidly, but the innovation process is becoming more efficient, faster and more flexible. At the same time, its complexity is increasing as more actors become involved. Consequently, innovation has to be seen not simply as a cross-functional process, but as a multi-institutional networking process.

This gives rise to the idea of a 'triple helix' – the complex web of relationships between universities, industries and governments, which characterizes late twentieth-century science and technology. Governments no longer attempt to 'orchestrate the construction of the knowledge infrastructure', but recognize the need for negotiation between a variety of players with differing needs and resources (Etzkowitz and Leydesdorff 1997). Giesecke

(2000) suggests that the triple helix model pictures innovation as a spiral movement that captures reciprocal relationships among institutional actors (public, private and academic) at different stages in the capitalization of knowledge. As a consequence, it changes the character of knowledge-producing institutions, and, we would add, the resultant direction of the technology trajectory.

All this has a non-spatial viewpoint. However, many commentators argue that geographic clustering of firms and related institutions has always been an important factor in economic development and that this is likely to continue (Scott 1998). Such clusters are envisaged as transaction-intensive regional economies, which are caught up in structures of interdependency stretching across the entire globe. They are important foundations of much contemporary international trade. Their importance is actually increased by the advent of advanced information and communications technologies, which have the effect of creating greater geographic differentiation, specialization and inter-regional trade.

Combining these institutional and geographic viewpoints suggests that there is an important role for agents of collective order, whether they be government, associations or public–private consortia (Scott 1998). Scott argues that economic competitiveness and growth can be much improved by policies that impact on the regional production system in three areas: critical inputs and services supplied as public goods, co-operation among firms in the tasks of production, and forums for strategic choice and action. Other commentators have drawn up recipes for governmental intervention that have more explicit forms of financial support (Giese 1987; Sternberg 1996). In a similar vein, but drawing more on the view that high technology industries are dependent on knowledge rather than on production networks, Finegold (1999) defines clusters as 'self-sustaining high-skill ecosystems' with characteristics that make them similar to natural ecosystems. However, institutional arrangements can create a 'supportive environment' for sustainability to take place. This is especially true of emerging high technology industries – such as opto-electronics.

Opto-electronics – technology and industry development

Opto-electronics (also called 'photonics') is one of a new breed of science-based technologies which involves manipulating materials at the atomic level (Kaounides 1995). Opto-electronics itself brings together two basic technologies (optics and electronics) in a process of 'technology fusion' (Dubarle and Verie 1993) and is recognized as a strategically important new high technology sector (JTEC 1996; NRC 1998). Miyazaki (1995) characterizes the industry in a three-level model which illustrates the idea of technology fusion, by showing how different end-use products evolve from the combination of intermediate products and underlying generic technologies.

Opto-electronics technologies emerged from a period of laboratory-based

R&D in the 1960s and 1970s (with the invention of the laser (Townes 1999) and fibre optics (Senior and Ray 1990)), into an applications and diffusion phase in the 1980s and 1990s. At this point, relationships between firms and infrastructure elements (such as research institutions, the labour market and government agencies) become more important, as R&D and skills are externalized into markets rather than hierarchies (Williamson 1975). The industry becomes characterized by the presence of large numbers of small medium-sized enterprises (SMEs), often in the form of networks surrounding the original players (Photonics Spectra 1995).

Opto-electronics in the UK and Germany

The growth of opto-electronics in the UK has been due largely to the impetus provided by the military and telecommunications markets, and the leadership of British Telecommunications (BT) and the Ministry of Defence (MoD) in fostering research in universities and private industry (ACOST 1988). BT was one of the first companies in the world to install optical fibres in its network. This pushed the technology into applications far quicker in the UK than in most other countries. The MoD instigated many new developments in night vision systems and liquid crystal displays. At the same time, there has been considerable inward investment in the UK. For example, Nortel Networks took over STC and then made Paignton a world centre for its manufacture of opto-electronic components. Other examples are Hewlett Packard taking over the ex-BT facility spun-off at Ipswich, and the Corning-BICC joint venture to produce optical fibre on Deeside.

The opto-electronics industry in Germany, by contrast, stems predominantly from developments in the optical and precision engineering sectors, rather than the electronics and communications industries. This contrast between radical innovation in the UK in newly emerging technologies and high-quality incremental innovation based on existing industries in Germany has been noted by Porter (1990) and Soskice (1997). Thus existing competences in optical and precision engineering were enhanced by the development of skills and knowledge in opto-electronics in large companies such as Carl Zeiss and Siemens, as well as within the research institutes, especially the Fraunhofer and Max Planck institutes, universities and polytechnics (Fachhochschulen).

Government support for opto-electronics in the UK

In the UK there are three main UK governmental institutions relevant to opto-electronics – the Department of Trade and Industry (DTI), the Engineering and Physical Science Research Council (EPSRC) and the Defence Evaluation Research Agency (DERA); for a more detailed account of the involvement of these institutions in the history of opto-electronics in the UK see Hendry *et al.* 2002. In summary, the picture is one of early support for scientific research followed by greater attention to the practical applications of the new technology.

We illustrate this process by looking at the development of the generic technology Metal Organic Vapour Phase Epitaxy (MOVPE), which is a process for producing opto-electronic wafers (in some ways comparable to silicon wafers in the semi-conductor industry). These wafers are one of the building blocks for opto-electronics, as they provide the source of diode laser light for a range of products.

1 MOVPE technology grew out of the silicon semiconductor industry and was developed in research laboratories in the UK, Germany and the USA (including the UK's MoD research centre DERA), to improve wafer fabrication technologies for materials needed for advanced opto-electronics applications.
2 The early work to commercialize this involved significant public–private collaboration. The MoD worked jointly with a UK-based telecommunications device manufacturer on early production capability for defence applications, while similar work went on at BT laboratories into more commercial opportunities.
3 Growing demand from the fibre-optic communications market for semi-conductor lasers based on these materials reinforced the need to develop the technology.
4 The Joint Opto-electronics Research Scheme (JOERS) in the early 1980s was an early example of the UK government intervening in this industry, and providing public money for joint academic–industry research to extend the understanding of MOVPE (and other technologies in opto-electronics). JOERS was a DTI scheme.
5 Equipment manufacturers developed a business in the reactors required to make semiconductor materials, initially at the R&D level, but later at production volumes. This technology subsequently moved out of the UK to Germany, which became the leading country for the supply of such reactors.
6 Much of the production capability went into large manufacturing organizations, but newer, smaller companies started up as specialist 'merchant' suppliers of epitaxy wafers. The LINK scheme, which followed JOERS from the DTI, supported a number of collaborative research projects into the application of MOVPE wafers.
7 Increasingly, MOVPE has become the standard technology for the fabrication of materials required for an expanded set of products, including high-brightness displays, which is one of the fastest growing segments of the industry.

This case example of the commercialization of MOVPE illustrates the need for collaborative activity among firms and universities in the UK, supported by government research contracts, in developing a technology that is very specialized and which has diverse applications.

Government support for opto-electronics in Germany

The relevant ministry for technology policy in Germany is the Federal Ministry of Education, Science, Research and Technology (BMBF), which is the obvious counterpart to the DTI. Other sponsors that are relevant to opto-electronics include the German Research Council and the German Space Agency, which play broadly similar roles to EPSRC and DERA in the UK (Hendry *et al.* 2002). However, the significant additional feature of the German technology infrastructure is the large number of (semi-) public research institutes, such as the Large Research Institutes, the Max-Planck Society and the Fraunhofer Society, as well as the universities. We illustrate the German system of government support for opto-electronics by looking at Laser 2000 as an example case. Laser 2000, which is 50 per cent funded by the BMBF, has as its objective the aim of securing a supply of reliable high-power laser diodes for system manufacturers and manufacturing applications.

1. German firms have traditionally been major players in the world market for industrial lasers, which conventionally depend upon gas laser technology (Strass 1996). This strong emphasis on manufacturing applications was influenced by early industrial innovators in the field such as Lambda Physik (Gottingen) and Rofin-Sinar (Hamburg).
2. However, in the 1990s, diode lasers, which are less powerful and typically used in telecommunications and optical data storage applications, were gaining in importance as new laser sources for heavier duty manufacturing applications.
3. The Laser 2000 initiative started in 1993. It had two major objectives: to develop new generations of diode laser equipment, and to open up new fields of application (BMBF 1995).
4. Altogether, some thirty partners were involved in the first phase of Laser 2000. This list included Fraunhofer (and other) institutes, universities, large firms and SMEs. In order to encourage the diffusion of the technology, BMBF set up a network of eleven 'try-out and advice centres in laser technology'. There was a particular focus on increasing laser applications among SMEs.
5. In 1998 BMBF announced an extension to Laser 2000, based around two main projects. One focused on R&D into advanced high-power diode lasers, while the other had a remit to explore new applications as well as optimizing the use of existing technology. The former consisted mainly of research institutions including the Fraunhofer-IAF at Freiburg as co-ordinator, but the latter primarily of industrial firms led by Robin-Sifar, and fourteen other firms and institutions.
6. One consequence of this focus on lasers is that Germany now has thirty-five institutes at universities and polytechnics, six Fraunhofer institutes, three Max Planck institutes, one Large Research Institute and nineteen

other research institutes active in the field of laser technology. This adds up to an extensive research infrastructure which covers nearly all parts of laser technology (BMBF 1995), although the BMBF also notes deficiencies in basic research.

In summary, we note in both cases the importance of the market context for the research projects and the need for collaborative effort between universities and large firms (and in the case of Germany, the intermediate research institutions). Both industries have clearly been extensively supported by government funding. However, in the UK there was an earlier emphasis on fundamental research into new technologies for emerging applications in contrast to the German approach, which focused more around existing industry strengths and discovering how new technologies could be used to improve performance. However, the later LINK programme in the UK was also more focused on applications.

Regional policy and initiatives

Regional initiatives in Wales

Wales has a history of dependence on coal and steel industries, which have now declined almost to the point of disappearance. For many years, starting in the 1930s, initiatives aimed at economic regeneration in selected niche markets have been tried out, the most lasting manifestations of which are the industrial estates developed to attract new industries to the area under the 1934 Special Areas Act. These became the platform for the successful policy of encouraging new inward investment.

Geographically, the region falls clearly into two halves: north and south. South Wales has the main legacy of the heavy industrial period, with a predominance of large urban areas and extensive industrial complexes. The South benefited most from inward investment and this is reflected in the nature of the opto-electronics firms, almost all of which are present precisely because of this factor. As a result, the picture is quite diverse, with no linking thread connecting the companies, nor is there any single large company acting in a co-ordinating role. There appears to be very little horizontal collaboration and the vertical supply chain connections are largely with overseas firms.

In the North the picture is very different. This is a rural area with hardly any indigenous heavy industry, with the notable exception of the extreme northeast corner where the steel industry once had a major presence. In one sense, the origin of the opto-electronics industry in North Wales is similar to that in the South, in that both began with inward investment. But the significant difference in the North is that it started earlier and its subsequent development has relied heavily on the growth trajectory of one company. In 1957, Pilkington plc took the decision to establish a new division in North Wales specializing in the production of high-quality ophthalmic glass (Barker 1994). From this initial investment, reinforced by a policy of diversification into new business areas, Pilkington spawned a number of independent companies, while retaining a core competence in advanced applied optics for military applications.

The main institutions in Wales relevant to opto-electronics are the Welsh Office and the Welsh Development Agency (WDA). In 1964 the Welsh Office was formed to take on responsibility for public administration in Wales. Its main role is policy development and a clear indication of its strategy for technology development is contained in *Pathway to Prosperity* (Welsh Office 1998: 24), a document that was published after an extensive consultation process. In a section on 'Sectoral balance', the document reads,

> We have concluded that policy should not concentrate on the promotion of particular sectors, but instead should be focused on correcting or removing the market failures which prevent industries from achieving their full potential. In particular we see a role for policy in developing and maintaining mutually supportive networks which will help companies grow.

The WDA was created in 1976 as an executive agency with one part of its brief being to bring new companies into Wales and to stimulate entrepreneurship. In the first of these it has been very successful, but while traditional location attributes have until now given Wales a comparative advantage, they are no longer sufficient to attract and retain high-quality investment. Wales (it is argued) needs a broader approach. Drawing on the view that firms learn best from other firms and that other firms are the most credible and effective tutors, the WDA has put much emphasis on designing and brokering inter-firm networks. Initiatives cover three separate fields: supplier development, technology support and training consortia.

The technology support programme, for example, aims to promote the generation and diffusion of new technologies, especially to SMEs. This has three distinct aspects: technology audits delivered by WDA staff, promotion of university centres of expertise and creation of technology clubs. These aim to integrate the hitherto diffused expertise of key sectors by encouraging more interaction between companies, research centres and government. The concept was pioneered by the Welsh Medical Technology Forum and has now been extended to other key sectors, including opto-electronics where the Welsh Opto-electronics Forum (WOF) was formed in 1996. The WOF has been an active force in bringing together a wide range of people with interests in the development of opto-electronics in Wales. As such, this is a rather unique, regionally focused, sector-specific network, distinct from trade associations in which firms normally participate and professional associations for individuals. A recent development has been a proposal to create a research institute in North Wales focusing on opto-electronics.

Regional initiatives in Thuringia

Thuringia has a long history as an industrial region with expertise in the automobile, optical, mechanical and, later, electronic industries. Geographically this specialization in terms of opto-electronics is in the so-called 'technology triangle' the district between three cities: Erfurt (microelectronics), Jena (optics, opto-electronics, chemistry and manufacturing technology), and Ilmenau (with its university focusing on technology).

A further feature important for opto-electronics was the existence in Jena of the 'Kombinat Carl Zeiss', which had extensive experience in many industry sectors dealing with optics and electronics. In the former East German state, a Kombinat was an integration of all sorts of research activities together with development and production facilities in one organization. Thus there were close relationships with university (and other) research centres, which in effect were part of the organization. After the reunification of Germany the Treuhand took over such institutions with the objective of transferring them into market capable units. But such a Kombinat did not fit capitalist patterns and had therefore to be restructured. In the case of Carl Zeiss, this restructuring resulted in two companies; Carl Zeiss Jena GmbH (which is linked with Zeiss Oberkochen in what was West Germany) and Jenoptik AG (which contained all the remaining companies in one integrated formation). Carl Zeiss remained more or less intact, retaining its traditional focus on optics and optical instruments. Jenoptik was left with the more diversified and unpromising part and it became the main focus of development attention.

Soon after the founding of Jenoptik, the former Prime Minister of the state of Baden-Württemberg, Lothar Späth, took on the job of CEO with a brief to make the company profitable. The problem was that Jenoptik did not have any products established in key markets, nor did it have a brand name that was known. Furthermore, the former major market, Eastern Europe, had faded away. The approach was twofold. First, it was crucial to focus on the development of new products and markets. This was done by concentrating on building up promising divisions. Second, for the other divisions, the plan was to disintegrate them into SMEs in order to encourage companies outside the region to come in and invest in new firms or merge with the disintegrated ones. This has turned out to be very successful and 200 firms have been attracted to the region (Späth 1999).

A second aspect of the recent history of Thuringia concerns universities and their staff. Many university members had to leave their organization after reunification, partly because of being involved in the previous political system and partly because their departments were made redundant by the restructuring. Some of the former departments became small laboratories specializing in high-technology tasks (e.g. measuring). Such laboratories nowadays play an important part in the network of high-technology companies.

A clear institutional indication of the value given to networks is OptoNet, which comprises forty-five organizations in opto-electronics. In 1999 this association of companies was founded with the objective of influencing the development of university education and occupational training. Since most of the companies operating in opto-electronics are SMEs, this is difficult for them to achieve on their own. A second aim of OptoNet is to establish Thuringia as a leader in Europe for opto-electronic systems technology. OptoNet has many features in common with WOF, but it is interesting to note that their priorities are reversed in that WOF puts the main emphasis on global connections and is now coming round to the idea of creating a learning resource.

Two Thuringian ministries (i.e. the Ministry of Science and Arts, and the Ministry of Economic Affairs and Infrastructure) are involved in the planning and implementation of indigenous technology policy and the implementation of federal schemes. In 1993, they commissioned a panel of experts from industry and science from Thuringia and outside to recommend in which direction the state should steer research and development (Hassink 1996). This 'Sörbe Commission' advised the government to focus Thuringian and federal support on future-oriented technologies in order to develop regional industry-science clusters. Two of the four technologies selected as a result for support are closely connected to opto-electronics (SFTT 1994).

Within the Ministry of Economic Affairs the small Department of Technology Support is mainly concerned with supporting technology in companies and fostering technology transfer along generic lines without giving particular preference to any one technology. Technological support for SMEs in Thuringia is of three kinds: support for technology in individual companies; advice and management support, and support for the acquisition and use of patent rights. Thuringia has its own local patent office, the Patentinformationszentrum und Online-Dienste (PATON). PATON is a department of the Technological University of Ilmenau, and is the official information centre on patents in the state of Thuringia as well as the official patent acceptance office. Again, there is the contrast with the WDA approach, which is more concerned with promoting technology transfer and diffusion.

Besides these technological aid schemes, the Thuringian government has set up a dense infrastructure of technology transfer agencies (including, for example, Thüringer Agentur für Technologietransfer und Innovationsförderung GmbH (THATI) which was founded in 1992), and incubator centres. The most successful example of the latter is the Technologie und Innovationspark Jena (TIP) (Scherzinger 1996). TIP offers tenants assessments of business plans, financial consultancy, advice on technological aid schemes, and refers firms to specialists they may need (such as business and tax consultancies). In many respects it operates like similar incubator centres in the UK.

How companies innovate

The facilities, resources, policies and initiatives supplied by government, at regional and national level, however, need to be set against the experience of companies themselves innovating. In this section, we provide an initial look (Tables 9.2 and 9.3) at the experience of companies in Wales and Thuringia. The analysis takes into account four factors: (1) the position of the company's technology in the three-level model of Miyazaki (1995), as this fundamentally affects its product market strategy; (2) the ownership picture, as this is an important dynamic in determining the purpose of innovation; (3) the source of innovation in terms of local and international technology-push and market-pull factors, and (4) the impact of innovation and its subsequent development.

Position in the three-level model

Most companies in both regions operate at the components and systems level in the Miyazaki three-level model. The major exception to this rule is Company WA in Wales, which has established itself as a leading manufacturer of custom wafers for the semiconductor device industry. This company illustrates the success of the inward investment policy both in encouraging new firm creation and in developing a global outlook. The two founders were heavily involved in the technology both as academics and industrialists before setting up in Wales. The company could have located anywhere in the UK but took advantage of regional assistance grants, support from the local government authority in setting up the plant, and WDA help in creating purpose-built premises on a new business park.

Moving above the generic technologies layer of the model, the distinction between components and systems is sometimes difficult to judge. Unless a company can establish a mass market for standard products, this position in the product market spectrum cannot usually be maintained and most companies aim to move up the value chain to supply key components and systems that are differentiated to meet their customer requirements. In the case of Wales, the main markets addressed are military (with commercial variants), communications and consumer products (the latter being due to the influence of company WB which came to Wales purely by chance). These markets reflect the early involvement of the industry with pioneering developments. In contrast, in Thuringia, the markets addressed are more industrial and scientific, based on sensing and analytical technologies that rely heavily on local expertise strongly grounded in research institutes and technology transfer initiatives.

Company ownership

The Welsh companies show a greater element of foreign ownership, with all bar two (WG and WH) having extensive involvement with foreign owners

Table 9.2 Summary of cases in Wales

Start year	*Position in three-level model*	*Ownership*	*Source of innovation*	*Developments*
WA (1987)	Generic technologies. Epitaxial wafers. Almost 100% sales to major OE component manufacturers outside UK.	Initially private. Now merged with US company and quoted on EASDAQ. Company started operations in Wales attracted by inward investment support.	Company was set up to commercialize production of MOVPE wafers based on experience in university and industrial research.	Extensive collaboration with research institutions both local and remote. Collaboration with key (German) supplier of manufacturing equipment.
WB (1991)	Systems. Production systems for making 'master discs' in optical disc industry.	Private after unsuccessful period of venture capital support. Company relocated to Wales for private reasons.	Founders built on previous experience making vinyl discs. Major breakthroughs into optical discs came after collaborative experience with MNCs in media industry.	Current generation equipment based on use of lasers. Now exploring product enhancement options based on electron beams.
WC (1971)	Components. Manufacture of pre-recorded optical discs.	Spin-off from WB originally. Now part of UK-based MNC with extensive interests in media products.	Product improvement to enhance appearance of disc and provide greater security resulted from approach by non-local company specializing in holographics.	Product improvement incorporated in new process equipment as part of factory expansion plans.
WD (1966)	Key optical components and systems for military markets. Image enhancement (night-sight) and display systems are core technologies.	Joint venture between UK glass manufacturer and French military hardware supplier. Originally part of UK company.	Optics expertise built up in military applications from green-field start in Wales in 1966. Major new product innovation derived from holographic research in parent UK company. Manufacturing expertise developed in-house.	Potential for holographic idea to be exploited in commercial vehicles. Group expansion opens up new market opportunities but concerns expressed about source of research inputs.

WE (1986)	Systems based on use of ruby and CO_2 lasers.	Private company status restored after unsuccessful merger with US company. Original start-up created by academic from local university.	Idea for product with mass-market potential came out of university research. Not a commercial success, mainly to do with flawed marketing arrangements.	Company is now going back towards the role of a commercial R&D organization. Looking to exploit local university research as customized solutions for industrial clients.
WF (1982)	Systems for optical imaging and delivery. Similar to WD but with a much stronger commercial bias.	Originally a spin-off from WD. After several ownership episodes is now part of a MNC. Close connections to a sister company in same group, based outside Wales.	Original expertise derived from WD, but now extended by inputs from sister company in group. Major new product development being driven by key customer.	Inclusion in large MNC now offers this company the chance to become the specialist centre for precision optics within the group.
WG (1989)	Laser marking systems for military and commercial aircraft.	Spin-off from UK defence contractor now quoted on OFEX.	Technology was discovered in former parent. WG set up to commercialize it.	Now looking for technology that will improve product and reduce cost.
WH (1993)	Components and systems for local area networks.	Originally a spin-off from WD who had diversified into optical fibre production and then retracted. Set up as an MBO.	Derived from WD but developed into cable and component manufacture rather than optical fibre.	Company is moving up the value chain into supplying complete systems and installation and service management.

Table 9.3 Summary of cases in Thuringia

Start year	*Position in three-level model*	*Ownership*	*Source of innovation*	*Developments*
TA (1990)	Analytical instruments (60%) and components (40%) for environment, medical and agricultural markets.	Private with substantial part (25%) owned by Jenoptik.	History of expertise in instrumentation in Carl Zeiss now moderated by key customers.	Extensive research collaboration with research institutions (not necessarily local). Expanding into micro and bio measurement systems.
TB (1993)	Prototype systems for industrial and research markets.	Consortium of 30 regional SMEs.	Research and development centre mainly in sensors and micro systems.	Initially dependent on local markets but now looking to expand outside region with own products.
TC (1991)	Systems for positioning and security applications.	MBO from Carl Zeiss.	One product idea with worldwide potential came from medical physician. Local customers initiate customized product ideas.	Exploit medical product globally. Expand customer base for industrial and meteorological products outside region.
TD (1995)	One-third each generic technologies, components and systems.	Jenoptik.	Diverse set of technologies built on optics and lasers developed in Carl Zeiss era for sensing and measuring applications in industrial and military markets.	Strategic intent to move up the three-level model and concentrate on products and systems. Use made of local research infrastructure but international sales now 60% means a greater focus on extra-regional sources of innovation.

TE (1993)	Components (lasers).	MBO from Carl Zeiss. Company HQ, sales and service in Munich. Jena is manufacturing location.	Company started as distribution outlet (in Munich) which then purchased gas laser facility in Carl Zeiss.	Traditional strength in gas lasers now being extended into newer types of laser, partly by acquisition.
TF (1990)	Components (optical elements and coatings) for industrial and scientific markets.	Private. Founder was academic at FSU Jena.	Founder's expertise in optics.	Initially dependent on local markets (still 50%) but now expanding outside region. Part of local network of research experience.
TG (1992)	Products and systems. Solutions based on integrated circuits with optical and electronic elements for sensing and signal processing applications. Industrial and IT markets.	Ownership split between Carl Zeiss, Jenoptik, two major customers, and management. 'Business-oriented research institution'.	Spin-off unit from Carl Zeiss. Carl Zeiss and Jenoptik still important customers.	50% of business is local, but now looking outside region and Germany. Local research collaboration with IOF.
TH (1992)	Key components. Application-specific integrated circuits (ASICs) for industrial applications.	Ownership changed twice after Treuhand. Now part of Belgian MNC.	Company goes back to 1937 as part of Telefunken. Major IC centre (with fabrication facilities) in GDR.	Specialist supplier of ICs to 'mixed signal (analogue to digital and HF) markets'. Extensive research collaboration intra- and extra-regional.

and collaborators or ownership by a UK multinational company (MNC) with extensive overseas interests. In Thuringia, the ownership pattern is much more locally based and structurally varied, with the participation of banks, co-operative associations, and Carl Zeiss and Jenoptik. This complex system of local ownership may be viewed as a geographically limiting factor, but other evidence suggests that Thuringian companies do have international connections. Carl Zeiss is a major company with strong export interests, and Jenoptik, which has recently been floated on the Frankfurt exchange, has made it clear that its ambitions lie in becoming an international concern by taking over internationally active companies that fit in with its own technologies. At the same time, Jenoptik regards co-operation with the local technological infrastructure and other companies in the region as a basis for this expansionist strategy. This particular aspect of the difference between the two regions may be due simply to the fact that the Welsh companies are older and have therefore been exposed to foreign influences for a longer time.

It is notable that both regions have examples of what in Thuringia are called 'business-oriented research institutions'. Table 9.1 shows that there are twenty of these companies in Thuringia, and two of them have a speciality in opto-electronics and feature as case studies in this chapter. Both of these companies are owned by consortia. TB is owned by some thirty regional SMEs, and TG is owned by Carl Zeiss, Jenoptik and several others, and specializes in contract research and development. These organizations have extensive links into the local research infrastructure and in their early company life relied heavily on local markets. In Wales, the WDA does not make a report on the number of such companies, but our research revealed one in opto-electronics. This company (WE) is a spin-off from a university research department, is privately owned and has already made one (unsuccessful) venture into international markets in collaboration with an American partner.

Source of innovation: (local) technology-push or (international) market-pull

It is commonly asserted that the capacity for innovation in a national economy is both derived from and embedded within the structures that support it. Such structures include public and private funding arrangements, education–industry links, and basic and applied research institutes (Harding 1999). As a result of historical development, these structures differ between countries, and this in turn leads to differing growth rates and differing technological competencies. From a transaction cost perspective, Nooteboom (1999) generalizes to two generic kinds of innovation system, which he typifies as European (and in particular German) and Anglo-American. In the Anglo-American model, innovation derives from a search for low transaction costs that are achieved by bargaining under the threat of using alternative partners. He suggests that this mode exhibits network relations between firms that are less durable, but have greater flexibility in configuration, with the result that Schumpeterian 'novel combinations' and radical innovations can flourish. Another effect is that there is likely to be a greater propensity for international collaboration. In contrast, the European style is more consensual, co-operative and locally networked, and likely to lead to incremental innovation. It should be noted that in the Nooteboom (1999) model, a crucial issue is the extent to which technology is flexible (capable of being able to generate differentiated products) or inflexible (in that it locks companies into specific investments and low-cost standard products). We assume that opto-electronics is a flexible technology and therefore more suited to the European co-operative style of innovation than the low transaction cost approach of the Anglo-American.

The evidence from our case studies goes some way to support the Nooteboom view. In the case of Wales, the key company is WD, whose technology derives from working with major defence contracting companies in the development of radical new imaging systems for military aircraft pilots. When this technology was extended and improved by the use of holographic techniques, WD drew on the research resources of its parent company, which in Nooteboom's view is another characteristic of the Anglo-American model (the prevalence of large integrated companies). Other examples from Wales reveal the importance to this regional innovation system of international alliances in such cases as WB, which developed optical disc production capability in collaboration with foreign MNCs, and WG, which derived its technology from a large UK defence contractor and then separated from it to market the new product on a global scale.

In the case of Thuringia, the evidence from recent years suggests that the approach is much more locally networked, consensual and informal, and results in incremental innovation. It is of course true that the history of opto-electronics in Thuringia is very much related to the fortunes of Carl Zeiss Jena, which, during the post-Second World War years, when all the important discoveries in opto-electronics were being made, was an important part of the military–technology complex in the Eastern Bloc countries. Without examining this period in greater depth, it is difficult to determine the consequences of this experience other than to note that at reunification, dramatic steps had to be taken to establish it in a Western context. Nevertheless, the presence of scientific and engineering expertise in large quantities has formed the basis of regeneration of the region, leading to incremental developments, grounded in technologies that were historically strong in the area, and shaped mainly by the market demands of local customers.

Developments

A key issue for both regions is the question 'where do we go from here?' Both regions now have established strong positions in opto-electronics, but as the pace of technological innovation quickens, keeping up to date is becoming more difficult as new ideas increasingly come from outside the native industrial setting. On the other hand, an important element of technology strategy has become not so much how to keep up as how to put existing technology to the best possible use. Accordingly, a balance needs to be drawn between exploration and exploitation, and this may be the key question for companies rather than globalization versus local development.

The indications from Tables 9.2 and 9.3 are that companies in Wales are now looking at ways in which they can consolidate their position by incorporating product and process improvements, whereas in Thuringia the emphasis is much more on expanding market opportunities outside the region. In Wales, two good examples of this are companies WB and WC which are both looking at including new technologies into their products and manufacturing processes in order to give them an advantage in the market. This fits with the Utterback (1994) view of industrial innovation that once products have reached a 'dominant design' stage the focus for innovation switches to process improvements. In both cases the development involves partners outside the region and technologies that are new to the companies. Kodama (1992) sees this as a process of 'technology fusion' and argues that this will be more sustainable than a strategy which looks for technological breakthroughs.

From a policy perspective, there are a number of parallels between the two regions in terms of intention, with some differences in implementation that are probably determined by the nature of the regional innovation systems. Both regional governments have policy directions that build on existing strengths and competencies, avoid championing one technology but rather encourage a range of technologies by providing infrastructural and institutional support for technology transfer, and give prominence to companies setting up networking clubs and industrial clusters. In Thuringia, the reliance on traditional institutions as sources of technology is now being moderated by a need to encourage the growth of new companies and engage with national and international markets. In Wales, the early involvement of companies in the region with opto-electronics and the priority given to inward investment has created a greater reliance on international sources for technological developments, but the WDA is now putting much more emphasis on creating a technology transfer process that involves the universities in the region.

Finally, the balance between basic research funded directly or indirectly by government and its development into commercial products involves a sensitive balance. This balance has shifted considerably in recent years, with more pressure on public research for relevance and commercial application.

In the UK this forces universities to engage more with industry, which is not always a culturally comfortable and intellectually acceptable process, and there is a degree of tension among UK academics over the proper relationship (Hendry *et al.* 2000b). At the same time, industry is less able to 'free-ride' on basic research in government research centres or research funded by government in companies. Unfortunately, many companies themselves have concluded that basic research is not their responsibility, and have refocused their R&D departments to deliver outputs with more immediate commercial benefits. This could leave a dangerous hiatus, in which not enough new path-breaking science is taking place. This is a concern shared by many, and is reflected in some of our cases.

In Germany, federal support for a complex institutional infrastructure, which helps a broad spectrum of companies respond to international competition by generating the skills and technology resources needed to pursue quality strategies, has always been regarded as a strength. However, since the 1992 to 1993 recession, this mode of intervention has been criticized within Germany for allegedly creating a competitiveness problem for companies. Critics claim (among other things) that the research system is too weak in radical innovation, with damaging consequences for the development of high-technology industries. The introduction of the 'American model' is advocated in order to encourage innovation, particularly in high technology (Vitols 1997). While Vitols does not accept this criticism, there is other evidence that small companies in emerging high technology areas such as (opto-) electronics, have relatively few links with intermediate research institutes in Germany (the same is true of the UK, but this was never regarded as a UK strength) (Mason and Wagner 1999).

Discussion and conclusion

The central question in this chapter was the extent to which national (and by extension regional) systems of innovation have been reduced in influence in a technological domain that is highly scientific and international in its very nature.

It is interesting to note that theories on national systems of innovation developed as a reaction to an abstract neoclassical view of the world, in which 'knowledge is assumed to float freely ... and all agents will draw techniques from the same global book of blueprints' (Lundvall 1998). National differences in culture and institutions may affect the way things are done, but it is assumed that the impact on the allocation of resources and the efficiency of production is so minor that they can be neglected. Lundvall (1998) argues that if one assumes that an essential factor in the analysis of economic success is the process of innovation, rather than the allocation of resources, then it can be seen why national systems and styles matter. This imposes fundamental uncertainty in the theory, as innovation is by definition a process where all alternative outcomes cannot be known in advance. If one also adds the reasonable assumption that competence and skills are unevenly distributed between individuals, organizations, regions and nations, this introduces learning skills and competence into the analysis. One intention behind the concept of national systems of innovation is thus to change the analytical perspective away from allocation to innovation and from making choices to learning.

Although it is not explicitly stated, a key principle underlying Rothwell's (1992) model of innovation is the attempt companies make to limit the consequences of technical and economic uncertainty. This is indicated in the model by the inclusion of feedback loops and iterative processes, and the need for interaction and consultation with outside parties. In more sophisticated versions, computer modelling and simulation of possible futures is further evidence of this. A further factor may be face-to-face contact at certain key stages in the process, in order to improve communication in the exchange and creation of new knowledge. This provides the link to geography in that the nature of the innovation process involving face-to-face interaction suggests strong links to location. The characteristics of innovation tend to make some aspects of technological activity locally confined (tacit knowledge, know-how, production processes), while other technological activities are more global in character (basic scientific research, scientific publication, patents) (Porter and Sölvell 1998). The extent to which the local factor needs to be fostered to complement increased use of electronic systems is unclear. The demise of large in-house research facilities forces firms to look elsewhere. But it may also break an important link in the innovation chain, in the need for close proximity at certain stages.

> I have a major concern on this. We have gone through a period of talking about concurrent engineering – that is, the importance of having the manufacturing guy sat next to the designer. In many companies such as ourselves, which operate at the systems, component and module level, systems are getting so complicated nowadays that it is almost impossible for the systems designer to see all the possible implications of his design in terms of tolerances and feasibility at the components level. The way that we used to handle this problem was the designer would walk on the shop-floor and figure out what was possible and incorporate it into his design. When you are operating from internationally separated sites or even regionally separated sites, that becomes more difficult. I do not see information technology helping. All our sites are now linked by video conferencing facilities, which we use extensively, and this is regarded as critical for maintaining contact. But you cannot get across the kind of things that I'm talking about.
>
> (Interview company WD)

This poses a serious issue for firms that are physically dispersed and for inter-firm and other relationships involving technology transfer. Electronic interchange may be very helpful where what is exchanged is codified data within established commercial relationships. It may also work in disseminating basic ideas (although the long time lag before new ideas are developed into commercial applications belies this). The real problem is the engineering or developmental work to create a viable product and commercial production process (cf. Yli-Kauhaluoma, Chapter 8, this volume). This is why most of our firms emphasize their in-house capabilities to do this. But sometimes the solution to a problem comes from being able to draw on new technologies or old ones from an unexpected source. The cluster idea and emphasis on locality may offer a substitute for the large hierarchical firm, where 'tacit knowledge' may be shared through direct interaction.

Note

1 The authors are grateful to the Anglo-German Foundation for the Study of Industrial Society for the financial support provided to this project.

References

ACOST (1988) *Opto Electronics: Building on our Investment*, London: HMSO.

Amin, A. and Thrift, N. (1992) 'Neo-Marshallian nodes in global networks', *International Journal of Urban and Regional Research* 16, 4: 571–587.

Barker, T.C. (1994) *Pilkington: An Age of Glass*, London: Boxtree.

Becattini, G. (1990) 'The Marshallian industrial district as a socio-economic notion', in F. Pyke, G. Becattini and W. Sengenberger (eds) *Industrial Districts and Inter-firm Co-operation in Italy*, Geneva: International Institute for Labour Studies.

BMBF (1995) *Laser 2000: Forderungskonzept 1993–1997*, Bonn: Bundesministerium für Bildung, Wissenschaft, Forschung und Technologie.

Casper, S. and Vitols, S. (1997) 'The German model in the 1990s: problems and prospects', *Industry and Innovation* 4, 1: 1–13.

Cooke, P., Davies, S., Kilper, H., Morris, J., Plake, R. and Wood, G. (1995) *Revitalising Older Industrial Regions: North-Rhine Westphalia and Wales Contrasted*, London: Anglo-German Foundation.

Cooke, P., Uranga, M.G. and Etxebarria, G. (1997) 'Regional innovation systems: institutional and organisational dimensions', *Research Policy* 26, 4–5: 475–491.

Dosi, G. (1982) 'Technological paradigms and technological trajectories', *Research Policy* 11, 3: 147–162.

Dubarle, P. and Verie, C. (1993) *Technology Fusion: A Path to Innovation. The Case of Opto-electronics,* Paris: Organization for Economic Co-operation and Development.

Etzkowitz, H. and Leydesdorff, L. (eds) (1997) *Universities in the Global Economy: A Triple Helix of University–Industry–Government Relations*, London: Cassell Academic.

Finegold, D. (1999) 'Creating self-sustaining, high-skill ecosystems', *Oxford Review of Economic Policy* 15, 1: 60–81.

Freeman, C. (1974) *The Economics of Industrial Innovation*, London: Penguin.

Ganter, H-D. (1997) 'Structural change in the German clothing industry and the reaction of firms'. Paper presented at the Thirteenth EGOS Colloquium, Budapest, July.

Garnsey, E. and Cannon-Brookes, A. (1993) 'Small high technology firms in an era of rapid change: evidence from Cambridge', *Local Economy* 7, 4: 318–333.

Giese, E. (1987) 'The demand for innovation-oriented regional policy in the Federal Republic of Germany: origins, aims, policy tools and prospects of realisation', in J.F. Brotchie, P. Hall and P.W. Newton (eds) *The Spatial Impact of Technological Change*, London: Croom Helm.

Giesecke, S. (2000) 'The contrasting roles of government in the development of biotechnology industry in the US and Germany', *Research Policy* 29, 2: 205–223.

Hahn, R. and Gaiser, A. (1994) 'Aktuelle innovationsprozesse im berarbeitenden gewerbe', *Zeitschrift fur Wirtschaftsgeographie* 38: 60–75.

Harding, R. (1999) 'One best way? The case of Germany in an era of 'globalisation'. Paper presented at the Fifteenth EGOS Colloquium, Warwick Business School, July.

Hassink, R. (1996) 'Regional technology policies in the old and new länder in Germany', *European Urban and Regional Studies* 3, 4: 287–303.

Hendry, C. and Brown, J.E. (1999) 'Clustering and performance in the UK opto-electronics industry'. Seventh High Technology Small Firms Conference, Manchester Business School.

Hendry, C., Brown, J.E. and DeFillippi, R. (2000a) 'Regional clustering of high technology-based firms: opto-electronics in three countries', *Regional Studies* 34, 2: 129–144.

Hendry, C., Brown, J.E. and DeFillippi, R. (2000b) 'Understanding the relationships between universities and high technology-based SMEs: the case of opto-electronics', *International Journal of Innovation Management* 4: 1.

Hendry, C., Brown, J.E., DeFillippi, R. and Hassink, R. (1999) 'Industry clusters as commercial, knowledge and institutional networks: opto-electronics in six regions in the UK, USA and Germany', in A. Grandori (ed.) *Interfirm Networks: Organization and Industrial Competitiveness*, London: Routledge.

Hendry, C., Brown, J.E., Ganter, H.D. and Hilland, S. (2002) *Understanding Innovation: How Firms Innovate and what Governments can do to Help: Opto-electronics in Wales and Thuringi*a, London: Anglo-German Foundation for the Study of Industrial Society.

JTEC (1996) *Optoelectronics in Japan and the United States*, Baltimore, MD: Japanese Technology Evaluation Centre.

Kaounides, L.C. (1995) *Advanced Materials*, London: *Financial Times*.

Keeble, D.E. (1994) 'Regional influences and policy in new technology-based firm creation and growth', in R. Oakey (ed.) *New Technology-based Firms in the 1990s*, London: Paul Chapman.

Kodama, F. (1992) 'Technology fusion and the new R&D', *Harvard Business Review* 70, 4: 70–78.

Lundvall, B. (1998) 'Why study national systems and national styles of innovation', *Technology Analysis and Strategic Management* 10, 4: 407–421.

Mason, G. and Wagner, K. (1999) 'Knowledge transfer and innovation in Germany and Britain: "intermediate institution" models of knowledge transfer under strain?', *Industry and Innovation* 6, 1: 85–109.

Miyazaki, K. (1995) *Building Competences in the Firm: Lessons from Japanese and European Optoelectronics*, London: Macmillan.

Nooteboom, B. (1999) 'Innovation and inter-firm linkages: new implications for policy', *Research Policy* 28, 8: 793–805.

NRC (1998) *Harnessing Light: Optical Science and Engineering for the 21st Century*, Washington, DC: National Research Council.

Photonics Spectra (1995) *The Photonics Corporate Guide to Profiles and Addresses*, Pittsfield, MA: Laurin.

Porter, M.E. (1990) *The Competitive Advantage of Nations*, London: Macmillan.

Porter, M.E. and Sölvell, Ö. (1998) 'The role of geography in the process of innovation and the sustainable competitive advantage of firms', in A.D. Chandler, P. Hagstrom and Ö. Sölvell (eds) *The Dynamic Firm; The Role of Technology, Strategy, Organization, and Regions*, Oxford: Oxford University Press.

Rothwell, R. (1992) 'Successful industrial innovation: critical factors for the 1990s', *R&D Management* 22, 3: 221–239.

Scherzinger, A. (1996) 'Forschung und entwicklung in den Ostdeutschen agglomerationen Jena und Dresden', *DIW-Vierteljahreshefte zur Wirtschaftsforschung* 2: 172–189.

Scott, A.J. (1993) *Technopolis: High-technology Industry and Regional Development in Southern California*, Berkeley and Los Angeles: University of California Press.

Scott, A.J. (1998) 'The geographic foundations of industrial performance', in A.D. Chandler, P. Hagstrom and Ö. Sölvell (eds) *The Dynamic Firm*, Oxford: Oxford University Press.

Senior, J.M. and Ray, T.E. (1990) 'Optical-fibre communications: the formation of technological strategies in the UK and USA', *International Journal of Technology Management* 5, 1: 71–88.

SFTT (1994) *Forschung und Technologie in Thuringen*, Erfurt: Thuringen Ministerium für Wissenschaft und Kunst.

Soskice, D. (1997) 'German technology policy, innovation, and national institutional frameworks', *Industry and Innovation* 4, 1: 75–96.

Späth, L. (1999) 'Transition from an industrial to a knowledge-based economy: the role of the private sector', in OECD (ed.) *Economic and Cultural Transition Towards a Learning City: The Case of Jena*, Paris: OECD.

Staber, U. (1996) 'Accounting for variations in the performance of industrial districts: the case of Baden Württemberg', *International Journal of Urban and Regional Research* 20, 2: 299–316.

Sternberg, R. (1996) 'Reasons for the genesis of high-tech regions – theoretical explanation and empirical evidence', *Geoforum* 27, 2: 205–223.

Strass, A. (1996) 'Germany's laser production hits all-time high', *Opto and Laser Europe* 33: 8.

Townes, C.H. (1999) *How the Laser Happened: Adventures of a Scientist*, Oxford: Oxford University Press.

Utterback, J.U. (1994) *Mastering the Dynamics of Innovation*, Boston, MA: Harvard Business School Press.

Vitols, S. (1997) 'German industrial policy: an overview', *Industry and Innovation* 4: 1.

Welsh Office (1998) *Pathway to Prosperity: A New Economic Agenda for Wales*, Cardiff: Welsh Office.

Williamson, O. (1975) *Markets and Hierarchies*, New York: Free Press.

10 'We have crossed the rubicon'

A 'one-team' approach to information technology[1]

David Knights and Darren McCabe

Introduction

Both in the literature and in practice, technology has tended to be conceived as a determinant or driver of the business in private sector organizations. Partly this is a reflection of a human preoccupation with finding simple causal explanations for complex realities (e.g. Blauner 1964; Woodward 1958, 1965; Thompson 1967). However, it is also a function of new or changing technologies appearing to be more easily captured than broader ranging organizational or social change. Criticisms of such approaches have been vociferous in recent years (MacKenzie and Wajcman 1985; Knights and Murray 1994; Bloomfield *et al.* 1996; Grint and Woolgar 1997; see also McLoughlin and Dawson, Chapter 2, and Laurila and Preece, Chapter 1, this volume). Yet just when we think that technological determinism has finally been buried, it rises from the ashes, most recently in the guise of business process re-engineering (see Hammer and Champy, 1993). Technological panaceas appear irresistible despite the irrefutable conclusion that they promise more than they can deliver.

For the uninitiated, the 'technology' debate in academic discourse has become increasingly esoteric as it has moved away from the technological or socio-technical systems determinism of the 1950s and 1960s. Attempts have been made to incorporate the 'social' into our understanding through various theoretical developments in the literature including gender studies (Cockburn 1983, 1985), labour process analysis (Noble 1984), social constructionism (Bijker *et al.* 1987), social shaping (MacKenzie and Wajcman 1985), organizational politics (Markus and Bjorn-Andersen 1987; Drory and Romm 1990; Knights and Murray 1994) and actor-network and the social science of knowledge perspectives (Callon *et al.* 1986; Latour 1987). None the less, management consultants and some academics rein in such advances through their one-dimensional prescriptions, and the 'techies' within organizations reinforce these tendencies towards simple technological determinist perceptions that ultimately inform the work of practitioners. Not least this is because comparatively rigid or dualistic distinctions between the 'social' and the 'technological' fit more easily with practitioners' everyday commonsense understandings of the world. They are also more compatible with management's conventional understanding of science as a method of seeking determinate causes and this, in turn, fuels an insatiable demand for simple or 'quick-fix' solutions to problems. Buying in technology and especially information and communication technology (ICT) is readily seen as such a 'quick fix'. Technology is viewed almost like a magic wand that assists in the 'mastering of "reality"' (Lyotard 1984: 47) seemingly without the aggravation of political tensions and contentions that are routinely encompassed in the social organization of work. It is largely for this reason that new ICTs, in the form of systems or electronic networks facilitated by the Internet, are presently seen to be in the driving seat of economic, social and political change. This chapter seeks to challenge the 'closure on reality' to which this view gives expression not so much by focusing on why it is wrong as illustrating, through an empirical case study, some of the implications of believing it to be right.

Although we agree with Grint and Woolgar (1997) that 'the boundary between the social and the technical is part of the phenomenon to be investigated' (p. 37; quoted in Laurila and Preece, Chapter 1, this volume) such distinctions are less important to us than exploring how the assumptions regarding, and the consequences of, a given technology lead innovations so frequently to be washed up on the shore of failed interventions. More importantly, we would want to consider the impact of working with particular technologies for the everyday lives of people who are subjected to them, or indeed, the thinking and power relations that go into subjecting others to technological innovations. Thus attempts to define what is technological and what is not seems to us to be of less importance than analysing the use of technology in practice.[2]

For this reason we see no need to locate the root of technology as a 'hard place' (McLoughlin and Dawson, Chapter 2, this volume) for this cannot avoid the slippery slope of increasingly esoteric and polarized epistemological disputes about material versus social reality (see Note 2). The self-defeating nuances of this debate are all too apparent in the discussion regarding whether technologies eventually stabilize or not. This is clear from McLoughlin and Dawson's own argument that 'we do not have to reject the concept of stabilization . . . but need to render it as a more contingent, malleable and iterative conceptualization of the obduracy of technology'. We would simply ask: Does it matter? We are not being flippant but believe that saying that a technology is not stable or, more appropriately, cannot be interpreted outside of its social context is no different from stating that it remains contingent, malleable and iterative. We are simply dealing with different ways of talking about the world that tend to pass each other by, often due, we believe, to fears of relativism and nihilism on the part of more traditional scholars. Thus to question the asocial black box of technology is not, for us, to say that anything goes. It is, however, to adopt a critical position by problematizing the often taken-for-granted assumptions underlying so many representations of the world. This is not to argue for an indifference to the way technology is represented but rather to keep our focus on the conditions and consequences of the use of these representations in everyday practices. In the research that forms the basis for this chapter, we found that information technology was represented as a panacea in the form of the One-Team Information Services (OTIS). This was expected to shift the organization out of the nineteenth (quill pens) into the twenty-first century where everyone was to communicate with each other electronically. However, when other more politically and economically sensitive issues arose, this panacea lost its priority, only to be replaced by another technology-driven transformation – the pursuit of e-commerce status. Since the burst of the dotcom bubble, we suspect that the e-commerce panacea will also have lost its overarching priority.

This chapter proceeds directly into an analysis of our case study since the Introduction to this book provides a review of the technology literature that is not in need of duplication. Throughout our presentation of the case study material, however, we seek to draw out various theoretical interpretations that make some reference to the prevailing literature on technology and organizational change. The focus is primarily on the introduction and then sudden demise of management change initiative, the OTIS. We seek to show how the faith in technology as a determinant of change in organizational practices and working arrangements is deeply problematic. Finally we draw out some comments in a discussion and conclusion.

The case study

During 1998, over a twelve-month period, interviews were conducted with thirteen middle to senior managers at Loanco's (pseudonym) Head Office. The main objective of these interviews was to gain an overview of the corporate strategy and change programme and to provide the context for a more detailed study of a call centre with the pseudonym Salesco. Within Salesco, twenty-one team members, amounting to one-third of the Salesco workforce, were interviewed. In addition, four team leaders and all four managers were interviewed for an hour. In January 2000 Loanco was revisited and we interviewed the Head of IT, a senior member of the team who introduced OTIS, we re-interviewed the Human Resources Director and also a senior HR Manager. In the case study we also draw on various company reports, team briefs, tape-recorded presentations, and video-recorded meetings and presentations. The approach was broadly ethnographical whereby we sought to grasp something of the way of life within Loanco. In this instance, we report upon our interpretation of the introduction of OTIS within the company.

The company

Loanco has a network of 250 branches and employs 3800 staff, 1000 of whom are based at its head office (HO). Its main area of business involves the provision, processing and maintenance of mortgage accounts. Strategic replacements in the senior management ranks in the late 1990s provided the impetus for a wide-ranging programme of change, including business process re-engineering. Moves are currently afoot to consolidate ten disparate processing units into a single centrally located processing centre. Tensions are already apparent insofar as Salesco staff expressed concerns regarding the staffing implications of re-engineering. For management, one way of responding to these concerns is for the company to expand in the future. Consequently, Loanco is pursuing an aggressive strategy of market growth, in conjunction with re-engineering, and so hopes to avoid large-scale redundancies.

Following his appointment in 1996, the new Chief Executive (CE) immediately set out a '100 day plan', which involved an extensive period of communication and consultation throughout the company. Roadshows were conducted in different areas of the country as a forum for communicating the corporate strategy of planned growth and the vision of Loanco as 'one team'. The Head of Re-engineering described the CE's approach during these roadshows:

> He gets on and talks about the need for change, and if you don't change you won't survive and that was what he described as his platform for change, his burning platform. A real need to drive us and he was very, I guess, inspirational in a way.

In conjunction with the roadshows, a video was prepared and made available to all staff. In it the CE introduced himself by elaborating aspects of his personal and employment background and his team-oriented philosophy on work, which he attributes to his experiences as a sportsman. In the video, the CE's closing comments were to ask staff to address him by his first name, and he expressed a wish that staff might allow him to address them by their first names. Clearly, such an open approach is unusual in financial services – staff appeared to warm to this and referred to the CE by his first name. Other symbolic changes which management has initiated more broadly in Loanco are abandoning preferential parking, toilet and restaurant facilities for managers, which Pam, the processing manager at Salesco considered to be 'very refreshing things to happen'.

While comparatively trivial, the CE's request that staff communicate with him on 'first-name terms' has meant that management is generally seen as more accessible than before. A less trivial development was that following consultation with the staff, management recognized that pay was a major area of dissatisfaction. Subsequently, a major pay review was initiated and in April 1998 all employees above the initial training grades were guaranteed a 3 per cent pay rise. None the less, change to date has not been without its contradictions and tensions. The CE has embarked on a series of innovations and strategic moves which are difficult to reconcile with the 'One Team' vision. These include a cost-cutting exercise at Head Office that was termed OVA or overhead value analysis that led to 300 redundancies and reductions in the number of branch staff. Furthermore, the CE has embarked on a series of corporate acquisitions that certainly promote group corporate growth but do little to secure the jobs of those engaged in the core function of providing mortgages.

The 'one-team' vision and strategy

> *We have crossed the Rubicon*. We are seriously committed to developing people . . . this is the toughest job to do. You cannot reverse a trend of under-investment overnight but by 2001 there will be real progress . . . that's the essence of this credo. . . . This is about 3800 people . . . *working together* to deliver on *a single vision* of the business. The key thing is that it's *one team* that is going towards 2001.
>
> (CE, 1997, Day 1, Strategy Conference; emphasis added)

The above extract is from a tape-recorded speech made in 1997 by the CE to his senior management team. Following a cascading approach, these managers (complete with transcribed copies of his speech) in turn delivered the vision and strategy to their staff. Providing a useful title for our chapter, the unitary message of 'working together' to achieve a 'single vision' is abundantly apparent in the messianic statement 'we have crossed the Rubicon'. Yet, it must also be understood as a command to those who are in doubt: culture as control (see Ray 1985; Willmott 1993). Both believers and non-believers alike must accept that things are different now. As we shall see, this 'one-team' approach is bound up with the CE's vision of transforming information services. However, his rhetoric may also be interpreted as a dividing practice (Foucault 1982) where anyone deviating from the discourse of unity will be immediately labelled as disruptive. In the speech the CE outlined his vision for the future and described his strategy under the acronym GEM: Growth, Efficiency and Managing Effectiveness. The company is to achieve growth and improved efficiency through managing effectively. Clearly, this vision is imbued with an emphasis upon performativity or the preoccupation with measuring the costs and benefits of work output. The CE also outlined what he referred to as 'the big stuff' which amounted to how the company is going to achieve this vision. One element was customer process re-engineering but there were also specific references to information technology:

> Let's look at the big stuff. At this stage all they [staff] need to know is that there is some serious big stuff going on . . . they need to get a sense of the commitment we have when they look at the big stuff. . . . There is

> a lot that has to be done on *efficiency* . . . common customer data access means one set of data at the front end for customers, we only enter data once . . . The *information network* that drives our business. We are *one step better than the quill*, at the moment . . . *the fact that we can't all communicate to each other electronically is an unsustainable position going forward.* (Emphasis added)

The substance of the CE's statement is the focus on technology. He outlined the view that improved information and communication technology is central to the company's future progress. In terms of technology, he described the company as being 'one step better than the quill' and this signifies that substantial investment and change is required. It should be clear from the evangelical way in which the CE speaks that he is a visionary; however, coercion lingered behind the inspirational rhetoric. Coercion is an undeniable element of re-engineering (Jackson 1996; Grint and Case 1998) but here we see it simultaneously permeating a much broader social vision of change that sought to encompass shared interests. Hence the CE continued his communication by saying:

> I expect you will have no trouble communicating to your colleagues that the goals are modestly stretching. I hope you will help them understand that growth equals opportunities and by the way no growth leads down a much less attractive path and *no efficiencies* leads down a much less attractive path. In fact it leads down a path that means *we haven't got a future.* So not only is it the best thing for them it's the only thing for us. (Emphasis added)

Thus, despite the rhetoric of the 'one team', we already see a contradiction as hierarchical control and power relations are reinforced through the vision. Control, especially over costs through efficiency gains, is firmly wedded to the unitary vision. The thinly veiled threat to staff of the failure to achieve growth also applies to the managerial audience. This vision is embedded in OTIS for, despite the rhetoric of the 'one team', profit, performance and productivity are the ends for which the new information service is the means. Yet it is not always possible to 'cost', particularly in the short term, either the benefits or the losses associated with a specific innovation (see Knights and McCabe 1997). Embedded as it is in the strategic vision, OTIS is not intended to pose a threat to the existing hierarchical power relations (though there is no guarantee of this). It must be able to demonstrate a substantial contribution to growth and efficiency. Thus a 'particular' social world is inscribed in the birth and genesis of OTIS (see Grint and Woolgar 1997) which is bound up with the demands of performativity. The following section will examine OTIS as it emerged and developed.

One-team information services

Traditionally Loanco was organized along functional lines that were referred to as 'silos' and these have generated a 'silo mentality'. Consequently, there are a number of distinct functions each having a separate 'subculture' that presents a considerable barrier to internal communications and knowledge sharing. In view of this, in October 1997, the CE organized a conference to which the key players from each functional group were invited. The Head of IT explained the rationale for the conference:

> Rather than just say things like lack of communication, we used a sort of knowledge theme as a way of getting people bought into the idea that there was a lot of knowledge in the business . . . there were some shared resources but not a great deal of them. And wouldn't it be a good idea if we could free up some of that knowledge and share it with other people.

Instead of confronting the participants of the conference with their limitations, there was a concern to engage them in designing a solution. The participants were involved in drawing up a picture of the business that went outside of the traditional organizational boundaries and this pointed towards the need for improved communication. It served as a springboard for OTIS. Hence, the delegates were prompted to identify that what was needed was a single method across the group for moving information and sharing knowledge. Two key elements of the new system would be a group Intranet and a single group-wide email. Although a number of systems were already in place they were often incompatible with one another.

In March 1998 the OTIS project began to be rolled out. The project had two objectives: first, to deliver benefits in terms of reduced costs, and second, to demonstrate that OTIS could be used to change not only the way people work but also their attitudes to work. The latter aim was made explicit in a corporate brief to all staff issued during May 1998:

> The aim for OTIS is that eventually we will *all be part of an online community*. With everyone swapping ideas and best practice, publishing information about their departments and linking to our customers and business partners directly. Working in a way which develops new relationships and *drives change in the society's culture* – in other words: the way we work, our attitudes and how we interact. (Emphasis added)

We can see then that the CE's corporate vision and strategy has had a major impact in terms of defining OTIS. Here the aims of OTIS are expressed in unitary terms with everyone being part of an 'online community'. It would seem that everyone is a potential beneficiary as they swap 'ideas and best practice'. The Head of IT explained that OTIS was not simply about 'technology implementation' as there was a concern to 'change the way people actually did their jobs'. None the less, there is an element of technological determinism here as it is believed that the introduction of OTIS will 'drive changes in the society's culture'. Thus it is apparent that the social and the technological are still considered to be distinct from each other, in this case the latter determining the former, although initially it was the opposite situation as the corporate vision set about preparing the 'social' for the 'technological': a case of social determinism (Gallie 1978). OTIS was firmly wedded to the need to cut costs, and this also reflects the corporate vision and strategy, as the brief continues:

> This pilot lays the firm foundations we need to grow OTIS into a thriving community. If the pilot is successful and identifies business benefits across the 'society' [i.e. the company] including *cost savings*, then the community will widen, and everyone can look forward to moving into OTIS. (Emphasis added)

Here we can see that the creation of a unitary, 'one-team' culture is not sufficient; there must be 'business benefits', 'including cost savings'. Yet what happened to the benefits of information sharing and communication in and of itself? The emphasis on 'business benefits' in terms of cost savings appears to override all other considerations including that of the 'team'. The 'bottom line' was clearly inscribed in the OTIS project. However, at the same time, management was conscious that culture and technological change can rarely be achieved if those who are its target are excluded from the process. Here we see management slipping away from technological determinism towards a kind of social determinism. But, as is often the case with new technology developments, it was difficult to secure the interest and involvement of staff initially, as few people actually understood anything about the Intranet. Although people eventually came up with ideas about how to use the Intranet it was not until they actually engaged with the technology and had learned how to use it that its relevance became apparent:

> We did spend a long time with a core group of people taking them around on visits and showing them what they were likely to get. To get to stimulate their imaginations, start to get them thinking about what that sort of technology could do for them. So they came up with ideas, initially, of how they might use it but we never really got to the nub of it until we'd given them the technology and the tools.
>
> (Head of IT)

Here is some clear evidence that the meaning of a technology becomes obvious only through its use and that any attempt to treat it independently of the social relations through which it is developed and applied is likely to result in failure (cf. Preece and Clarke, Chapter 3, this volume). Even with the best intentions the social ground cannot be prepared so as to be compatible with the technological, for what is technological is already social immediately it is made sense of (or not) in practical everyday affairs. While techies, consultants and the media tend to create representations of technology that are irrepressibly positive and 'progressive', in practice the benefits revolve around users finding the technology 'useful'.

In terms of the roll-out, various types of users were identified and different forms of technology and training were issued accordingly. The first group was defined as 'management' users. These are individuals who manage others and take decisions. Managers were provided with a mail client and a web browser. The second group of users was defined as 'knowledge workers'. These are described as specialists, individuals who have an analytical content to their job, who need information from a variety of sources and may therefore benefit from having access to the web. They are seen as information gatherers, disseminators, people who make recommendations, and are usually IT business analysts, legal staff and HR personnel. These individuals have a powerful, full-blown client on their desktop. The third group of users are defined as 'core workers' who are primarily customer-facing people. These members of staff require a PC and need access to information and mail via a browser, though their work does not involve a great deal of processing. Of these users, 25 per cent were management, 25 per cent were deemed to be knowledge workers and 50 per cent were core workers. We can see then that users are already clearly distinguished on hierarchical and functional grounds and this serves to reinforce earlier power relations, stratifications and divisions.

A small project team of twenty staff oversaw the pilot and engaged with users to determine how the Intranet should be structured, its appearance and content, and how it would be organized. A pre-eminent concern was that OTIS should 'deliver benefits' and these had to be tangible. This was a question that continually confronted users. Thus staff were asked to think about their everyday jobs, about the documents they produce, how they are produced and whether they could be put on the Intranet. Improvements in productivity of between 10 and 50 per cent were expected. It was believed that staff who spend a great deal of time communicating with other people, and also individuals on help desks who spend a huge amount of their time answering queries, could be freed up by putting information on to the Intranet, thereby reducing the number of phone calls. The staff were also asked to document their work both before and after OTIS so as to estimate the amount of time saved following its introduction. There was an attempt to list the benefits by identifying tasks that could be displaced by the Intranet such as producing reports, printing letters, using paper and making phone calls.

The bulk of the savings were in people's time but, according to the Head of IT, the problem was convincing management that the time saved could be used productively rather than simply 'wasted'. It became apparent that although staff were willing to adopt the technology and to change the way they worked, there was not a similar degree of support from management to enact the changes necessary to turn the time saved to other productive uses. The project was dogged by the intangible nature of the savings, leading to questions about how to make use of this spare 'time'. Management could either cut the number of staff employed or encourage staff to deploy their time more productively. The former is an unsavoury task for most people while the latter is incongruous with the culture of Loanco. As a number of managers pointed out, 'There is no tradition of cost benefit analysis in Loanco and in the past projects have simply been introduced rather than carefully costed.' Although a demand for cost savings and 'benefits', as outlined in the corporate vision, is embedded in the rationale for OTIS, this rationale proved difficult to operationalize. At the grass-roots level it seems that the managerial culture has not been touched by the corporate vision; consequently, management was loath to take advantage of the possible benefits OTIS had to offer. After all they were intangible benefits and management at all levels was seemingly unwilling to commit:

> To do this successfully means that you have to change the way that managers manage in this place. . . . So suddenly this becomes more than just implementing an intranet and a mail system.
>
> (Head of IT)

This suggests that even when it is endorsed by staff, the introduction of technology does not necessarily shift attitudes and behaviour that are deeply culturally embedded and entrenched. However, this is the case for management every bit as much as for staff. The pilot began to be rolled out in September 1998 and was completed by December 1998.

OTIS post-pilot scheme review

Following the pilot it was agreed initially that OTIS should be fully rolled out across the group. The key theme reported in a management summary of the pilot was that 'People want OTIS. They can see the future benefits and the way it could make communication easier.' In terms of the branch users, however, it was stated that:

> The culture of people being responsible for ensuring that they access information being provided and ensuring that they take care to keep themselves up to date by using the system has not been universally successful.
>
> (Management summary of pilot)

This is an interesting finding given the apparent support for OTIS. It suggests that staff may have considered OTIS to be an undue imposition upon their time. The report underlined the importance of committed local leaders to ensure that there is sufficient support for OTIS. These leaders 'must be advocates, enthusiasts, and *have time to make the most of their role* to maximize the benefits achieved' (emphasis added). Having the time to ensure that OTIS works effectively seems to be important even for enthusiasts and so we can see that there are decisions and choices to be made between the pressures on work output and for there to be sufficient time for effective communication and information sharing through OTIS. To date, work pressures seem to be taking priority over communication even among the local leaders for, as the management summary of the pilot continued, there were those 'who feel that it is another responsibility, which had no customer priority and defer from any ownership of the service being provided'. This reluctance among some local leaders to take OTIS on board supports our interpretation that for some individuals OTIS is seen as an additional burden.

A conference that included several workshops was conducted after the pilot as a post-pilot implementation review and a report of the conference findings was compiled in December 1998. It identified what people have done differently since the introduction of OTIS. Of the sixty-eight uses that were identified, the most popular were use of the calendar facility (thirteen), sending external emails (eleven), attaching documents (ten) and also making use of the electronic phone directory (nine). Only four respondents felt that there was a general improvement in communication while only two considered that they have shared knowledge as a consequence of OTIS.

During a benefits realization workshop, OTIS was found to have a number of 'business benefits problems' including 'identifying the benefits', 'difficult to cost benefits' and identifying 'who actually benefits?' This led to questions such as 'How do we measure the benefits as they cascade through the business?'; 'How can the saved time be used more profitably?' and 'Do we need to measure everything financially?' Despite the absence of a historical culture of cost–benefit analysis (reported above), we can see that OTIS reflected the performativity embedded in the corporate vision in terms of the preoccupation with efficiencies, control, cost-cutting and measuring benefits. Yet a substantial problem was to quantify the benefits of OTIS. One could argue that the preoccupation with cost cutting is antithetical to the concern to share information and improve communication, since the latter requires time, trust and co-operation. Moreover, it seems that identifying the financial benefits of OTIS will only lead to yet more cost-cutting measures, which both management and staff are keen to avoid. Some supposedly 'non-financial' or at least intangible benefits of OTIS were identified; these included:

- Aiding public relations of support functions.
- Ease of administration.
- Incremental savings of many small things.
- Openness is generating questions – there is a need to affirm what is 'the new way'.
- Debate and discussion should be encouraged.
- Better quality service.

These intangible benefits reflect the limitations of the strategic vision and strategy in that it reinforces a preoccupation with measuring the benefits of OTIS. In doing so, the vision excludes many issues that are potentially of great benefit to the company. Thus the vision and strategy are not only embedded in the rationale for OTIS but they are also its measure of success or failure, adoption or otherwise. It is intriguing that, on the one hand, the benefits of OTIS are self-evident (i.e. improved communication, shared knowledge) but on the other, such benefits only stand if they can be costed. At the post-pilot implementation review conference, the Human Resource Director reiterated the importance of OTIS and restated the link between OTIS and the corporate vision:

> Without fast, clear, appropriate communication it would be impossible to become 'the mutual which really can meet the needs of everyday customers throughout Britain'. (Conference report)

On the one hand then, there is a blind faith in technology, following a perception that technology, in and of itself, is a means to improve communication and to secure future competitive success. On the other hand, however, it seems that OTIS will be implemented only if it meets the imperative of delivering tangible savings. A closet technological determinism leads some managers down the path of thinking that IT can solve its problems. However, the intangible benefits and power relations that prevent its take-up, as outlined above, reveal the other ways in which OTIS is bound up with the social fabric of the company. In view of this, the benefits cannot be neatly defined as if they exist outside the social context of their application and use. Because measurement demands that technology as a variable be independent of its effects (dependent variables), the complex relations and benefits go unrecorded. In short, some of the problems of OTIS seem linked to a commonsense (epistemological) faith in causal analysis and quantification.

The way in which the costs/benefits of OTIS are bound up with the social context of the organization was apparent during an 'Internet-searching workshop'. The workshop report argued that 'the Internet should not be used as it is not part of the job'. A cost-based explanation for this comment is apparent in the 'concern over the time it takes to hunt for information'. Thus the miracle cure and benefits of OTIS were brought into question by the potential costs involved. Indeed, OTIS also appears to introduce uncertainties in that staff can develop new freedoms over which management has little control. Consequently, concerns were voiced with regard to 'searching' on the Internet as articulated in the following staff guidelines:

- There is no guarantee that the information you require is actually available or that when found it is correct.
- Initially you should restrict the amount of time that you 'surf' when trying to get information.
- The Intranet and Internet are only another source of information.

These guidelines provide direct evidence of how, despite the apparent intensification of surveillance made possible by ICT (e.g. Sewell and Wilkinson 1992), management is often unable to control the use and outcome of the technology itself (see Preece and Clark, Chapter 3, this volume). For not only is the technology difficult to cost, it also creates new space and scope for staff resistance and autonomy. Perhaps this is why in the OTIS pilot roll-out – feedback report it was stated that 'the promotion of the use of the service and the resolution of any issues should come from the top'. It seems then that management at all levels is anxious about committing itself to OTIS because neither its costs nor its benefits can be accurately accounted for and controlled.

OTIS: the aftermath

After the pilot for OTIS was complete, the company began to face intense pressure to transfer from being a mutual organization to a plc, and at this time OTIS was effectively shelved. Its demise was explained in terms of the need to conserve 'costs' or that it was a 'business decision'. Although approval to go ahead with full roll-out was not granted, the number of users was extended from 400 to 1000 without any concern to 'cost' this decision. The Head of IT explained this in terms of the CE's concern to communicate to as many members of staff as possible the importance of Loanco remaining a mutual organization. Indeed, remaining mutual had been a cornerstone of the CE's corporate vision and strategy. Here we can see how money is available providing that those in power deem that the issue is of sufficient importance.[3] In the haste to extend coverage of OTIS to as many staff as possible, the preoccupation with monitoring benefits was sidelined. Remaining mutual was more important than the costs involved.

Discussion and conclusion

We have examined the changes in Loanco without engaging heavily in the academic debates about the relationship between technology and the social. None the less, we have intimated a position that we would argue is different in its rejection of technological determinism from either the social constructionists or the social shaping theorists. Grint and Woolgar (1992: 376) argued that the technological is a function of the effort required to show that it is social. Like earlier polarized debates between realism and relativism, there is a danger here of degenerating into academic 'points scoring' over increasingly more obscure intellectual arguments as to what is technological and what is social (see Note 2). We have refrained from adding to this esoteric debate and concentrated more on the way that ideas about technology appear to be reproduced in management practice. Nor have we sought to denigrate the way in which managerial common sense returns persistently to some variant of technological determinism.

Rather, our intent has been to question the practical consequences of a managerial faith in technology. Of course, the ICT promoted by Loanco was embedded in the social since it was seen as both the content and consequence of everyone working as 'one team'. The technology – the Intranet – was seen to be the vehicle if not the stimulant for sharing knowledge in the company. As companies grow, they are dependent increasingly on 'managing at a distance'. Those responsible for direction and development in an organization require more information and knowledge to increase the 'efficiency of their representations' (Cooper 1992: 266) of the 'sharp end' of the business from which senior managers are necessarily remote. Without knowledge and information-rich representations of how the organization works, executive decisions could be entirely inconsistent or incompatible with routine practices. This becomes even more important as staff assume greater autonomy and authority (empowerment) – a trend reinforced by the more widespread use of ICT. But the political character of organizations makes neither the construction of appropriate representations nor the effective use of them a straightforward, let alone guaranteed, matter.

In the pilot of OTIS there is no question that opportunities arose for staff to use the Intranet in ways that saved time, for example, but there was little evidence that middle management was prepared to support this. One reason was that to cost these savings would require not only staff but also line and middle management to become more accountable and open to the gaze of surveillance. Such accountability was necessary in order to make use of the slack left in the system following improvements in knowledge sharing and communication. A number of uncertainties arose over whether such improvements would lead to redundancies or different ways of working. Moreover, it was not at all clear that management would allow sufficient time for staff to make use of OTIS taking into account the daily operational pressures of work. Thus it would seem that management and organizational politics itself is an obstacle to 'management by distance'. The inactivity was reinforced by an unwillingness among senior management to commit to OTIS when the savings were seemingly so intangible and the need to save costs was so imperative. OTIS was also unpredictable in that it offered staff new ways to escape control either by not using it or misusing it. In this way, OTIS was thwarted by the emphasis on performativity in the corporate vision and strategy.

The reason why OTIS was discontinued was also partly the failure of the IT manager to persuade other managers to change their practices. This was necessary if the time saved by staff adopting the new technology was to be redirected into productive channels. Clearly it is more difficult to change the behaviour of managers who have an equal or perhaps even more senior ranking than the person (i.e. the Head of IT) who is seeking to transform practices. Aside from these power relations, it seems to us that other developments had run ahead of 'this' management obstacle which had to take priority. The board of the company sought to resist pressures to demutualize and felt the need to demonstrate not only cost-cutting rationalizations but also 'progressive' developments that would impress the City equity fund managers. Rather than continue with a strategy that might be extremely beneficial internally but could not be guaranteed to reduce costs, the company refrained from committing itself to OTIS. Instead, it began to explore the possibilities of e-commerce.

In conclusion, it has to be argued that the politics of organization is a major factor in the development of organizational change, for it is representations of technology and the social that provide individuals and groups with the material artefacts and symbolic messages for mobilizing colleagues and networks in support of particular changes. These representations are drawn upon discursively and interpreted for purposes of securing a given outcome. The technology is not independent of the social but, as in our case study, is drawn into the service of other strategic objectives. The problem, however, is that following common-sense and everyday distinctions between the social and the technological, practitioners attempt to draw a line between the two. In doing so, management divorces technology, or in our case OTIS, from the relations of power and knowledge that are both its condition and consequence. It emphasizes performativity and 'formal rationality', and therefore the benefits or otherwise of any innovation such as OTIS are constructed in a highly restricted fashion. This in turn has a significant impact upon what management is able or willing to achieve and indeed, the possibilities of knowledge sharing and communication appear to have been lost in a mist of politics, culture, accountability, resistance, cost control, profit and efficiency.

Notes

1 We would like to acknowledge ESRC funding support from the Innovation Programme Grant Number L125251061.

2 The heated debate between 'realists' and 'constructionists' (sometimes denigrated as relativists) in relation to technology can readily run into an intellectual cul-de-sac. 'Self declared "realists" dream up a formulation which turns, so to speak, the "world upside down". They then challenge self-described "constructivists" to show how it can stay that way. This sets off an endless circle in which increasingly preposterous propositions are constructed (by realists) only to be "de-constructed" (by constructivists) so the whole thing can start all over again – until both sides lose interest' (Knights *et al.* 2002: 12). While, for example, the dispute over whether guns are as effective in warfare as roses is not uninteresting, few of the protagonists provide narratives of the conditions and contexts in which social enactments might favour one or the other (cf. ibid.)

3 This is not unlike the situation when governments go to war. Billions of dollars or pounds may be expended in a matter of days, whereas during peacetime money is seemingly unavailable for a host of worthy causes.

References

Bijker, W., Hughes, E., Thomas P. and Pinch, T.J. (eds) (1987) *The Social Construction of Technological Systems: New Directions in the Sociology and History of Technology*, Cambridge, MA: MIT Press.

Blauner, R. (1964) *Alienation and Freedom*, Chicago, IL: University of Chicago Press.

Bloomfield, B., Coombs, R., Knights, D. and Littler, D. (eds) (1996) *Information Technology and Organizations: Strategies, Networks and Integration*, Oxford: Oxford University Press.

Boudon, R. (1981) *The Logic of Social Action*, London: Routledge & Kegan Paul.

Callon, M., Law, J. and Rip, A. (1986) *Mapping the Dynamics of Science and Technology: Sociology of Science in the Real World*, London: Macmillan.

Cockburn, C. (1983) *Brothers: Male Dominance and Technological Change*, London: Pluto.

Cockburn, C. (1985) *Machinery of Dominance: Women, Men and Technical Know-How*, London: Pluto.

Cooper, R. (1992) 'Formal organization as representation: remote control, displacement and abbreviation', in M. Reed and M. Hughes (eds) *Rethinking Organization: New Directions in Organization Theory and Analysis*, London: Sage.

Drory, A. and Romm, T. (1990) 'The definition of organizational politics: a review', *Human Relations* 43, 11: 1133–1154.

Foucault, M. (1982) 'The subject and power', in H. Dreyfus and P. Rabinow (eds) *Michel Foucault: Beyond Structuralism and Hermaneutics*, New York: Harvester Press.

Gallie, D. (1978) *In Search of the New Working Class*, Cambridge: Cambridge University Press.

Grint, K. and Case, P. (1998) 'The violent rhetoric of reengineering: management consultancy on the offensive', *Journal of Management Studies* 35, 5: 557–577.

Grint, K. and Woolgar, S. (1992) 'Computers, guns and roses: what's social about being shot?', *Science, Technology and Human Values* 17, 3: 366–380.

Grint, K. and Woolgar, S. (1997) *The Machine at Work; Technology, Work and Organization*, Oxford: Polity Press.

Hammer, M. and Champy, J. (1993) *Reengineering the Corporation: A Manifesto for Business Revolution*, London: Nicholas Brealy.

Jackson, B.G. (1996) 'Re-engineering the sense of self: the management and the management guru', *Journal of Management Studies* 33, 5: 571–590.

Kling, R. (1991) 'Computerization and social transformations', *Science Technology and Human Values* 16, 3: 342–367.

Knights, D. and McCabe, D. (1997) 'How would you measure something like that? Quality in a retail bank', *Journal of Management Studies* 43, 3: 371–388.

Knights, D. and Murray, F. (1994) *Managers Divided*, London: Wiley.

Knights, D., Noble, F., Vurdubakis, T. and Willmott, H. (2002) 'Allegories of creative destruction: technology and organisation in narratives of the e-economy', in S. Woolgar (ed.) *Virtual Society?*, Oxford: Oxford University Press.

Latour, B. (1987) *Science in Action*, Milton Keynes: Open University Press.

Lyotard, F. (1984) *The Postmodern Condition*, Manchester: Manchester University Press.

MacKenzie, D. and Wajcman, J. (eds) (1985) 'Introduction' to *The Social Shaping of Technology*, Milton Keynes: Open University.

Markus, M.L. and Bjorn-Andersen, N. (1987) 'Power over users: its exercise by systems professionals', *Communications of the ACM* 30, 6: 498–504.

Noble, D. (1984) *Forces of Production: A Social History of Industrial Automation*, New York: Knopf.

Ray, A. (1985) 'Corporate culture: the last frontier of control?' *Journal of Management Studies* 23, 3: 287–297.

Sewell, G. and Wilkinson, B. (1992) 'Someone to watch over me: surveillance, discipline and the JIT labour process', *Sociology* 26, 2: 419–446.

Thompson, J.D. (1967) *Organizations in Action*, New York: McGraw-Hill.

Willmott, H. (1993) 'Strength is ignorance; slavery is freedom: managing culture in modern organizations', *Journal of Management Studies* 30, 4: 515–555.

Woodward, J. (1958) *Management and Technology*, London: HMSO.

Woodward, J. (1965) *Industrial Organization: Theory and Practice*, Oxford: Oxford University Press.

Index